BIOLOGICAL, SOCIAL, AND ORGANIZATIONAL COMPONENTS OF SUCCESS
FOR WOMEN IN ACADEMIC SCIENCE AND ENGINEERING

REPORT OF A WORKSHOP

Committee on Maximizing the Potential of Women in
Academic Science and Engineering

Committee on Science, Engineering, and Public Policy

NATIONAL ACADEMY OF SCIENCES,
NATIONAL ACADEMY OF ENGINEERING, AND
INSTITUTE OF MEDICINE
OF THE NATIONAL ACADEMIES

THE NATIONAL ACADEMIES PRESS
Washington, D.C.
www.nap.edu

THE NATIONAL ACADEMIES PRESS • 500 Fifth Street, N.W. • Washington, DC 20001

NOTICE: The project that is the subject of this report was approved by the Governing Board of the National Research Council, whose members are drawn from the Councils of the National Academy of Sciences, the National Academy of Engineering, and the Institute of Medicine. The members of the committee responsible for the report were chosen for their special competences and with regard for appropriate balance.

Support for this project was provided by the National Academies and the National Institutes of Health Office for Research on Women's Health under contract 1-OD-4-2137, task order 166. Any opinions, findings, conclusions, or recommendations expressed in this publication are those of the author(s) and do not necessarily reflect the views of the organizations or agencies that provided support for the project.

International Standard Book Number-10: 0-309-10041-0 (Book)
International Standard Book Number-13: 978-0-309-10041-0 (Book)
International Standard Book Number-10 0-309-65451-3 (PDF)
International Standard Book Number-13: 978-0-309-65451-7 (PDF)

Library of Congress Control Number 2006933601

Committee on Science, Engineering, and Public Policy, 500 Fifth Street NW, Washington, DC 20001; 202-334-2807; Internet, http://www.nationalacademies.org/cosepup.

Additional copies of this workshop summary are available from the National Academies Press, 500 Fifth Street NW, Lockbox 285, Washington, DC 20055; (800) 624-6242 or (202) 334-3313 (in the Washington metropolitan area); Internet, http://www.nap.edu.

Copyright 2006 by the National Academy of Sciences. All rights reserved.

Printed in the United States of America

THE NATIONAL ACADEMIES
Advisers to the Nation on Science, Engineering, and Medicine

The **National Academy of Sciences** is a private, nonprofit, self-perpetuating society of distinguished scholars engaged in scientific and engineering research, dedicated to the furtherance of science and technology and to their use for the general welfare. Upon the authority of the charter granted to it by the Congress in 1863, the Academy has a mandate that requires it to advise the federal government on scientific and technical matters. Dr. Ralph J. Cicerone is president of the National Academy of Sciences.

The **National Academy of Engineering** was established in 1964, under the charter of the National Academy of Sciences, as a parallel organization of outstanding engineers. It is autonomous in its administration and in the selection of its members, sharing with the National Academy of Sciences the responsibility for advising the federal government. The National Academy of Engineering also sponsors engineering programs aimed at meeting national needs, encourages education and research, and recognizes the superior achievements of engineers. Dr. Wm. A. Wulf is president of the National Academy of Engineering.

The **Institute of Medicine** was established in 1970 by the National Academy of Sciences to secure the services of eminent members of appropriate professions in the examination of policy matters pertaining to the health of the public. The Institute acts under the responsibility given to the National Academy of Sciences by its congressional charter to be an adviser to the federal government and, upon its own initiative, to identify issues of medical care, research, and education. Dr. Harvey V. Fineberg is president of the Institute of Medicine.

The **National Research Council** was organized by the National Academy of Sciences in 1916 to associate the broad community of science and technology with the Academy's purposes of furthering knowledge and advising the federal government. Functioning in accordance with general policies determined by the Academy, the Council has become the principal operating agency of both the National Academy of Sciences and the National Academy of Engineering in providing services to the government, the public, and the scientific and engineering communities. The Council is administered jointly by both Academies and the Institute of Medicine. Dr. Ralph J. Cicerone and Dr. Wm. A. Wulf are chair and vice chair, respectively, of the National Research Council.

www.national-academies.org

COMMITTEE ON MAXIMIZING THE POTENTIAL OF WOMEN IN ACADEMIC SCIENCE AND ENGINEERING

DONNA E. SHALALA [IOM], (*Chair*) President, University of Miami, Miami, Florida
ALICE M. AGOGINO [NAE], Roscoe and Elizabeth Hughes Professor of Mechanical Engineering, University of California, Berkeley, California
LOTTE BAILYN, Professor, Sloan School of Management, Massachusetts Institute of Technology, Cambridge, Massachusetts
ROBERT J. BIRGENEAU [NAS], Chancellor, University of California, Berkeley, California
ANA MARI CAUCE, Executive Vice Provost and Earl R. Carlson Professor of Psychology, University of Washington, Seattle, Washington
CATHERINE D. DEANGELIS [IOM], Editor-in-Chief, The Journal of the American Medical Association, New York, New York
DENICE DENTON,* (Deceased) Chancellor, University of California, Santa Cruz, California
BARBARA GROSZ, Higgins Professor of Natural Science, Division of Engineering and Applied Sciences, and Dean of Science, Radcliffe Institute for Advanced Study, Harvard University, Cambridge, Massachusetts
JO HANDELSMAN, Howard Hughes Medical Institute Professor, Department of Plant Pathology, University of Wisconsin, Madison, Wisconsin
NAN KEOHANE, President Emerita, Duke University, Durham, North Carolina
SHIRLEY MALCOM [NAS], Head, Directorate for Education and Human Resources Programs, American Association for the Advancement of Science, Washington, DC
GERALDINE RICHMOND, Richard M. and Patricia H. Noyes Professor, Department of Chemistry, University of Oregon, Eugene, Oregon
ALICE M. RIVLIN, Senior Fellow, Brookings Institution, Washington, DC
RUTH SIMMONS, President, Brown University, Providence, Rhode Island
ELIZABETH SPELKE [NAS], Berkman Professor of Psychology, Harvard University, Cambridge, Massachusetts
JOAN STEITZ [NAS], Sterling Professor of Molecular Biophysics and Biochemistry, Howard Hughes Medical Institute, Yale University School of Medicine, New Haven, Connecticut
ELAINE WEYUKER [NAE], Fellow, AT&T Laboratories, Florham Park, New Jersey

---

*Served from September 2005 to June 2006.

MARIA T. ZUBER [NAS], E. A. Griswold Professor of Geophysics, Massachusetts Institute of Technology, Cambridge, Massachusetts

Principal Project Staff

LAUREL L. HAAK, Study Director
JOHN SISLIN, Program Officer
BERYL BENDERLY, Consultant Science Writer
NORMAN GROSSBLATT, Senior Editor
JUDY GOSS, Senior Program Assistant
JENNIFER HOBIN, Christine Mirzayan Science and Technology Policy Graduate Fellow
RACHAEL SCHOLZ, Christine Mirzayan Science and Technology Policy Graduate Fellow
ERIN FRY, Christine Mirzayan Science and Technology Policy Graduate Fellow

COMMITTEE ON SCIENCE, ENGINEERING, AND PUBLIC POLICY

GEORGE WHITESIDES (*Chair*), Woodford L. and Ann A. Flowers University Professor, Harvard University, Boston, Massachusetts
UMA CHOWDHRY, Vice President, Central Research and Development, DuPont Company, Wilmington, Delaware
RALPH J. CICERONE (Ex officio), President, National Academy of Sciences, Washington, DC
R. JAMES COOK, Interim Dean, College of Agriculture and Home Economics, Washington State University, Pullman, Washington
HAILE DEBAS, Executive Director, UCSF Global Health Sciences, Maurice Galante Distinguished Professor of Surgery, San Francisco, California
HARVEY FINEBERG (Ex officio), President, Institute of Medicine, Washington, DC
MARYE ANNE FOX (Ex officio), Chancellor, University of California, San Diego, California
ELSA GARMIRE, Sydney E. Junkins Professor, School of Engineering, Dartmouth College, Hanover, New Hampshire
M.R.C. GREENWOOD (Ex officio), Professor, Nutrition and Internal Medicine, University of California, Davis, California
NANCY HOPKINS, Amgen Professor of Biology, Massachusetts Institute of Technology, Cambridge, Massachusetts
WILLIAM H. JOYCE (Ex officio), Chairman and CEO, Nalco, Naperville, Illinois
MARY-CLAIRE KING, American Cancer Society Professor of Medicine and Genetics, University of Washington, Seattle, Washington
W. CARL LINEBERGER, Professor of Chemistry, Joint Institute for Laboratory Astrophysics, University of Colorado, Boulder, Colorado
RICHARD A. MESERVE, President, Carnegie Institution of Washington, Washington, DC
ROBERT M. NEREM, Parker H. Petit Professor and Director, Institute for Bioengineering and Bioscience, Georgia Institute of Technology, Atlanta, Georgia
LAWRENCE T. PAPAY, Retired Sector Vice President for Integrated Solutions, Science Applications International Corporation, La Jolla, California
ANNE C. PETERSEN, President, Global Philanthropic Alliance, Kalamazoo, Michigan
CECIL PICKETT, President, Schering-Plough Research Institute, Kenilworth, New Jersey
EDWARD H. SHORTLIFFE, Professor and Chair, Department of Biomedical Informatics, Columbia University Medical Center, New York, New York

HUGO SONNENSCHEIN, Charles L. Hutchinson Distinguished Service Professor, Department of Economics, The University of Chicago, Chicago, Illinois
LYDIA THOMAS (Ex officio), President and Chief Executive Officer, Mitretek Systems, Inc., Falls Church, Virginia
SHEILA E. WIDNALL, Abby Rockefeller Mauze Professor of Aeronautics, Massachusetts Institute of Technology, Cambridge, Massachusetts
WM. A. WULF (Ex officio), President, National Academy of Engineering, Washington, DC
MARY LOU ZOBACK, Senior Research Scientist, Earthquake Hazards Team, US Geological Survey, Menlo Park, California

Staff

RICHARD BISSELL, Executive Director
DEBORAH D. STINE, Associate Director
LAUREL L. HAAK, Program Officer
MARION RAMSEY, Administrative Coordinator

Preface

Twenty-five years ago, Congress passed the Science and Engineering Equal Opportunity Act, which declares it "the policy of the United States that men and women have equal opportunity in education, training, and employment in scientific and technical fields." Major advances have occurred since then in the numbers of women enrolling in science and engineering classes in high school and college, but academic institutions are not fully using the growing pool of women scientists and engineering graduates that these classes have produced.

The nation's ability to use all its scientific talent is vital to its ability to retain technological and economic leadership in an increasingly competitive world. A diverse workforce brings new perspectives and priorities to science and engineering education and research. Removing artificial barriers that prevent scientists from making their optimal contributions therefore has high priority.

Over the last 40 years, the number of women studying science and engineering has increased dramatically. Women now earn 51% of the bachelor's degrees and 37% of PhDs, including 45% those in biomedical fields. Within the population of women science and engineering students, there are divergent experiences. For example, white women earn 50% of the bachelor's degrees and 41% of the PhDs awarded to whites. Hispanic women earn 55% of the bachelor's degrees and 50% of the PhDs awarded to Hispanics. African American women earn 64% of the bachelor's degrees and 54% of the PhDs awarded to African Americans.

Nevertheless, women do not hold academic faculty positions in numbers commensurate with their increasing share of the science and engineering talent pool. This is particularly true for African American women. The discrepancy exists at both the junior and senior faculty levels but is especially great at the top research-intensive universities. Furthermore, women who find academic employment are less likely than men to have tenure-track jobs in science or engineering departments or to advance to tenure. Even when they land tenure-track jobs and earn tenure, women lag behind men in salary, professional honors, and positions of authority.

The causes of the discrepancies are controversial. Observers have attributed differences in career progression and success to sex differences in cognitive abilities, to differences in career interests and preferences, to bias and discrimination, to gendered institutional policies and practices, to broader societal gender roles and assumptions, or to a combination of these factors.

To explore the question, the National Academies Committee on Science, Engineering, and Public Policy assembled the ad hoc Committee on Maximizing the Potential of Women in Academic Science and Engineering and charged it to

- Review and assess the research on sex and gender issues in science and engineering, including innate differences in cognition, implicit bias, and faculty diversity.
- Examine the institutional culture and practices of academic institutions that discourage and prevent talented individuals from realizing their full potential as scientists and engineers.
- Determine effective practices to ensure that women doctorates have access to a wide range of career opportunities in academe and in other research settings.
- Determine effective practices for recruiting and retention of women scientists and engineers in faculty positions.
- Provide recommendations to guide faculty, deans, department chairs, other university leaders, funding organizations, and government agencies in the best ways to maximize the potential of women science and engineering researchers.

As a vital part of its effort, the committee held a public convocation, Maximizing the Potential of Women in Academic Science and Engineering: Biological, Social and Organizational Components of Success, on December 9, 2005, in Washington, DC.[1] The convocation consisted of three elements: a series of panel discussions, poster sessions where attendees shared their data and experiences, and a public comment session. We brought together national experts in a number of disciplines to discuss crucial and controversial questions. Speakers were asked to address what sex differences research tells us about capability,

---

[1]The meeting agenda and speaker presentations are available online at http://www7.nationalacademies.org/womeninacademe/.

PREFACE *xi*

behavior, career decisions, and achievement; the role of organizational structures and institutional policy; cross-cutting issues of race and ethnicity; key research needs and experimental paradigms and tools; and the ramifications of their research for policy, particularly for evaluating current and potential academic faculty.

Speakers presented the most up-to-date research exploring the effects of sex and gender[2] on cognition and on recruiting, hiring, promoting, and retaining women scientists and engineers, and they described the best methods for improving women's opportunities to advance and succeed in academic science.

Although the discussions during those activities helped the committee to respond to its charge, this report presents the views and opinions of the convocation participants and may not reflect the views of the committee or of the National Academies. The committee released a final consensus report with findings and recommendations in September 2006.

> Donna E. Shalala, *Chair*
> Committee on Maximizing the
> Potential of Women in Academic
> Science and Engineering

---

[2]Sex is defined as "the biological state of being male or female" and gender as "the culturally prescribed characteristics and roles of a male or female in society and associated with masculinity and femininity."

Acknowledgments

The Committee on Science, Engineering, and Public Policy appreciates the support of the National Academies standing Committee on Women in Science and Engineering (CWSE), which is represented on the Guidance Group, on the study committee, and through staff support.

This report is the product of the efforts of many people. We would like to thank those who spoke at our convocation (in alphabetical order):

MAHZARIN RUSTUM BANAJI, Richard Clarke Cabot Professor of Social Ethics, Harvard University, and Carol K. Pforzheimer Professor, Radcliffe Institute for Advanced Study, Cambridge, Massachusetts

ROBERT DRAGO, Professor of Labor and Women's Studies, Pennsylvania State University, State College, Pennsylvania

SUSAN FISKE, Professor of Psychology, Princeton University, Princeton, New Jersey

JAY GIEDD, National Institute of Mental Health, National Institutes of Health, Bethesda, Maryland

DONNA GINTHER, Associate Professor of Economics, University of Kansas, Lawrence, Kansas

DIANE HALPERN, Professor and Chair of Psychology, Berger Institute for Work, Family, and Children, Claremont McKenna College, Claremont, California

JANET HYDE, Professor of Psychology and Women's Studies, University of Wisconsin, Madison, Wisconsin

JOANNE MARTIN, Fred H. Merrill Professor of Organizational Behavior, Graduate School of Business, Stanford University, Stanford, California

BRUCE McEWEN [NAS/IOM], Professor, The Rockefeller University, New York, New York

KELLEE NOONAN, Diversity Program Manager, Technical Career Path, Hewlett Packard, Sunnyvale, California

JOAN REEDE, Dean for Diversity and Community Partnership and Associate Professor of Medicine, Harvard Medical School, Cambridge, Massachusetts

SUE ROSSER, Professor and Dean, Ivan Allen College, Georgia Institute of Technology, Atlanta, Georgia

ANGELICA STACY, Professor, Department of Chemistry, University of California, Berkeley, California

JOAN WILLIAMS, Distinguished Professor of Law and Director, Center for WorkLife Law, University of California, Hastings College of the Law, San Francisco, California

YU XIE, Otis Dudley Duncan Professor of Sociology, University of Michigan, Ann Arbor, Michigan

This report has been reviewed in draft form by those selected for their knowledge, expertise, and wide range of perspectives, in accordance with procedures approved by the National Research Council's Report Review Committee. The purpose of this independent review is to provide candid and critical comments that will assist the institution in making the published report as sound as possible and to ensure that the report meets institutional standards of objectivity, evidence, and responsiveness to the study charge. The review comments and draft manuscript remain confidential to protect the integrity of the deliberative process. We thank the following for their participation in the review of this report:

ROBERT DRAGO, Professor of Labor Studies and Women's Studies, Pennsylvania State University, State College, Pennsylvania

EVELYNN HAMMONDS, Senior Vice Provost for Faculty Development and Diversity, Harvard University, Cambridge, Massachusetts

KRISTINA JOHNSON, Professor and Dean, Pratt School of Engineering, Duke University, Durham, North Carolina

JAMES C. KAUFMAN, Assistant Professor of Psychology, California State University at San Bernardino

JOANNE MARTIN, Fred H. Merrill Professor, Graduate School of Business, Stanford University, Stanford, California

CHERRY MURRAY, Deputy Director for Science, Lawrence Livermore Laboratory, Livermore, California

LONDA SCHIEBINGER, The John L. Hinds Professor of History of Science, Stanford University, Stanford, California

ABIGAIL STEWART, Professor of Psychology, University of Michigan, Ann Arbor, Michigan

ACKNOWLEDGMENTS

Although the reviewers had many constructive comments and suggestions about the report, they were not asked to endorse the findings and recommendations of the report, nor did they see a final draft of the report before its release. The report review was overseen by May R. Berenbaum, Professor and Head of the Department of Entomology at the University of Illinois Urbana-Champaign, appointed by the Report Review Committee, who was responsible for making certain that an independent examination of this report was carried out in accordance with institutional procedures and that all review comments were carefully considered. Responsibility for the final content of this report rests entirely with the authoring committee and the institution.

In addition, we thank the Guidance Group that oversaw this project:

NANCY HOPKINS (*Guidance Group Chair*), Amgen Professor of Biology, Massachusetts Institute of Technology, Cambridge, Massachusetts
ELSA GARMIRE, Professor, School of Engineering, Dartmouth College, Hanover, New Hampshire
W. CARL LINEBERGER, Professor of Chemistry, Joint Institute for Laboratory Astrophysics, University of Colorado, Boulder, Colorado
ANNE C. PETERSEN, President, Global Philanthropic Alliance, Kalamazoo, Michigan
MAXINE SINGER, President Emerita, Carnegie Institution of Washington, Washington, DC
HUGO SONNENSCHEIN, Charles L. Hutchinson Distinguished Service Professor, Department of Economics, The University of Chicago, Chicago, Illinois
LILLIAN WU, Director of University Relations, International Business Machines, New York, New York
MARY LOU ZOBACK, Senior Research Scientist, Earthquake Hazards Team, US Geological Survey, Menlo Park, California

Finally, we thank the staff of this project for their guidance, including Laurel Haak, program officer with the Committee on Science, Engineering and Public Policy and study director, who managed the project; John Sislin, the collaborating program officer from the Committee on Women in Science and Engineering; Beryl Benderly, the science writer for this report; Judy Goss, who provided project support; Christine Mirzayan Science and Technology Graduate Policy Fellows Jennifer Hobin, Rachael Scholz, and Erin Fry, who provided research and analytical support; Jong-On Hahm, former director of the Committee on Women in Science and Engineering; Peter Henderson, director of the Committee on Women in Science and Engineering; Mary Mattis, senior program officer, National Academy of Engineering; Richard Bissell, executive director and Charlotte Kuh, deputy executive director of the Policy and Global Affairs; and Deborah D. Stine, associate director, of the Committee on Science, Engineering, and Public Policy.

Contents

Introduction   1

Section 1: Summaries of Convocation Sessions   7
Panel 1: Cognitive and Biological Contributions
Panel Summary   10
    Gender Differences and Similarities in Abilities: Janet Hyde, Department of Psychology, University of Wisconsin, Madison, 11
    Sexual Dimorphism in the Developing Brain: Jay Giedd, National Institute of Mental Health, National Institutes of Health, 15
    Environment-Genetic Interactions in the Adult Brain: Effects of Stress on Learning: Bruce McEwen, The Rockefeller University, 17
    Biopsychosocial Contributions to Cognitive Performance: Diane Halpern, Berger Institute for Work, Family, and Children, Claremont McKenna College, 20
Selections from the Question and Answer Session, 24

Panel 2: Social Contributions   28

Panel Summary   29
    Implicit and Explicit Gender Discrimination: Mahzarin Rustum Banaji, Department of Psychology, Harvard University and Radcliffe Institute for Advanced Study, 30
    Contextual Influences on Performance: Toni Schmader, Department of Psychology, University of Arizona, 32

xvii

Interactions Between Power and Gender: Susan Fiske, Department of Psychology, Princeton University, 38
Social Influences on Science and Engineering Career Decisions: Yu Xie, Department of Sociology, University of Michigan, 42
Selections from the Question and Answer Session, 44

Panel 3: Organizational Structures   48

Panel Summary   49
Moving Beyond the "Chilly Climate" to a New Model for Spurring Organizational Change: Joan Williams, Center for WorkLife Law, University of California, Hastings College of the Law, 51
Economics of Gendered Distribution of Resources in Academe: Donna Ginther, Department of Economics, University of Kansas, 56
Bias Avoidance in the Academy: Challenges, Opportunities, and the Value of Policies: Robert Drago, Departments of Labor and Women's Studies, Pennsylvania State University, 61
Gendered Organizations: Scientists and Engineers in Universities and Corporations Joanne Martin, Graduate School of Business, Stanford University, 64
Selections from the Question and Answer Session, 69

Panel 4: Implementing Policies   72

Panel Summary   73
Recruitment Practices: Angelica Stacy, Department of Chemistry, University of California, Berkeley, 74
Reaching into Minority Populations: Joan Reede, Harvard Medical School, 81
Creating an Inclusive Work Environment: Sue Rosser, Ivan Allen College, Georgia Institute of Technology, 89
Successful Practices in Industry: Kellee Noonan, Technical Career Path, Hewlett Packard, 91
Selections from the Question and Answer Session, 93

Section 2: Workshop Papers   97
Donna Ginther, *The Economics of Gender Differences in Employment Outcomes in Academia,* 99
Diane Halpern, *Biopsychosocial Contributions to Cognitive Performance,* 113
Janet Shibley Hyde, *Women in Science and Mathematics: Gender Similarities in Abilities and Sociocultural Forces,* 127

Sue V. Rosser, *Creating an Inclusive Work Environment*, 137
Joan C. Williams, *Long Time No See: Why Are There Still So Few Women in Academic Science and Engineering?* 149
Yu Xie, *Social Influences on Science and Engineering Career Decisions,* 166

Section 3: Poster Abstracts   175
Sociology   177
Florence Bonner and Vernese Edgeh, *Policy and Praxis: Advancing Women in Higher Education and Influencing Outcomes*, 177
Miguel R. Olivas-Luján, Ann Gregory, John Miller, JoAnn Duffy, Suzy Fox, Terri Lituchy, Silvia Inés Monserrat, Betty Jane Punnett, and Neusa María Bastos F. Santos, *Successful Academic Women in the Americas: Human and Social Capital Descriptors,* 178
Gloria Scott, *Science Is Foundation for Leadership,* 180
Roberta Spalter-Roth, *Work-Family Policies in Academia as Resources or Rewards,* 180
Monica Young, *Case Studies from the Female Engineering Professoriate,* 181

Organizational Structure   182
Amber Barnato and Pamela Peele, *The Role of Informal Organizational Structures on Women in the Health Sciences,* 182
Diana Bilimoria, Susan R. Perry, Xiangfen Liang, Patricia Higgins, Eleanor P. Stoller, and Cyrus C. Taylor, *How Do Female and Male Faculty Members Construct Job Satisfaction?* 183
Diana Bilimoria, C. Greer Jordan, and Susan R. Perry, *A Good Place to Do Science: Creating and Sustaining a Productive, Inclusive Work Environment for Female and Male Scientists,* 183
Diana Bilimoria, Margaret M. Hopkins, Deborah A. O'Neil, and Susan R. Perry, *An Integrated Coaching and Mentoring Program for University Transformation,* 184
Cheryl Geisler, Deborah Kaminski, Robyn Berkley, and Linda Layne, *Up Against the Glass: Gender and Promotion at a Technological University,* 185
Rachel Ivie, *Women in Academic Physics and Astronomy*, 186
Mary Ellen Jackson, Phyllis Robinson, Sarah Conolly Hokenmaier, and J. Lynn Zimmer, *Faculty Horizons: Recruiting a Diverse Faculty,* 186
Delia Saenz and Allecia Reid, *Diversity in STEM Disciplines: The Case of Faculty Women of Color,* 187

Institutional Policy   188

    Ruth Dyer and Beth A. Montelone, *Initiatives to Increase Recruitment, Retention and Advancement of Women in Science and Engineering Disciplines at Kansas State University*, 188

    Lisa Frehill, Mary O'Connell, Elba Serrano, and Cecily Jeser-Cannavale, *Effective Practices for STEM Faculty Diversity*, 189

    Jo Handelsman, Molly Carnes, Jennifer Sheridan, Eve Fine, and Christine Pribbenow, *NSF ADVANCE at the UW-Madison: Three Success Stories*, 190

    Peggy Layne, Patricia Hyer, and Elizabeth Creamer, *Institutional Transformation at Virginia Tech*, 190

    Janet Malley, Pamela Raymond, and Abigail Stewart, *Institutional Transformation at the University of Michigan*, 191

    Nancy Martin, Beth Mitchneck, and William McCallum, *Scientifically Correct: Speaking to Scientists about Diversity*, 192

    Geralidine L. Richmond, *Working to Increase the Success of Women Scientists in Academia*, 192

    Eve A. Riskin, Kate Quinn, Joyce W. Yen, Sheila Edwards Lange, Suzanne Brainard, Ana Mari Cauce, and Denice D. Denton, *Leadership Workshops to Effect Cultural Change*, 193

    Tammy Smecker-Hane, Lisa Frehill, Priscilla Kehoe, Susan V. Bryant, Herb Killackey, and Debra Richardson, *ADVANCE: Successful Recruitment of Womento STEM at UCI*, 194

Section 4: Appendixes
A Workshop Agenda   197
B Speaker Biographical Information   202
C Committee Biographical Information   211
D Statement of Task   221

FIGURES, TABLES, AND BOXES

Figures

Section 1

1-1   Cross-cultural differences in fifth-grade mathematics performance, 14
1-2   Longitudinal development of white matter, 16
1-3   Biopsychosocial model, 22
1-4   Gender differences in mathematics performance, 34
1-5   Teaching about stereotype threat inoculates against its effects, 38

1-6 Fiske et al.'s Stereotype Content Model applied to subtypes of women, 41
1-7 Percentage of doctorates granted to females, 58
1-8 Percentage of tenured faculty who are women, 59
1-9 Women fast-track professionals with babies in the household, by age of professional, 62
1-10 Physical science, mathematics, and engineering applicant pool and faculty positions at The University of California, Berkeley, 76
1-11 Biological and health sciences applicant pool and faculty positions at the University of California, Berkeley, 77
1-12 Departmental hiring vs the applicant pool, University of California, Berkeley, 78
1-13 Children in households among assistant professors at the University of California, Berkeley, 80
1-14 Number of science and engineering bachelor's degrees awarded to minority females, by race and ethnicity, 1994-2001, 84
1-15 Number of science and engineering doctorates awarded to minority-group women, by race and ethnicity, 1994-2001, 85
1-16 Medical school faculty by rank, gender, race, and ethnicity, 86
1-17 Number of science and engineering doctorate holders employed in science and engineering occupations in universities and 4-year colleges, by race, ethnicity, and faculty rank, 2001, 87

Section 2

2-1 Percentage of doctorates granted to females, 1974-2004, 103
2-2 Percentage of tenured faculty who are female, by discipline, 1973-2001, 104
2-3 Gender differences in tenure track job within 5 years of PhD, 105
2-4 Gender differences in promotion to tenure 10 years past PhD, 105
2-5 Gender salary gap by academic rank, 2001 SDR, 109
2-6 Biopsychosocial model in which the nature-nurture dichotomy is replaced with a continuous feedback loop, 117
2-7 An example of a mental rotation task. Can the pairs of figures in A and B be rotated so that they are identical? Reaction times and correct answers are recorded, 119
2-8 Gender differences in achievement: 15 year old and 8th grade students, 122
2-9 Average SAT scores of entering college classes, 1967-2004, 123
2-10 Georgia Institute of Technology female faculty by rank and year, institution-wide, 145
2-11 Georgia Institute of Technology faculty flux charts, 146

2-12 Synthetic cohort life course, career processes, and outcomes examined, and data sources, 168
2-13 Sex-specific probabilities for selected pathways to an S/E baccalaureate, 170
2-14 Trends in female-male ratio of publication rate, 172

Tables

Section 1

1-1 Methods Used by University of California, Berkeley Departments to Enhance Faculty Hiring Pool, 79
1-2 Intentions of Freshman to Major in Science and Engineering Fields, by Race, Ethnicity, and Sex, 2002, 83

Section 2

2-1 The Magnitude of Gender Differences in Mathematics Performance as a Function of Age and Cognitive Level of the Test, 129
2-2 Effect Sizes for Gender Differences in Mathematics and Science Test Performance Across Countries, 133
2-3 Total Responses to Question 1, 140
2-4 Categorization of Question 1 across Year of Award, 141
2-5 Standardized Mean Gender Difference of Math Achievement Scores Among High School Seniors by Cohort, 169
2-6 Female-to-Male Ratio of the Odds of Achieving in the Top 5% of the Distribution of Math Achievement Test Scores Among High School Seniors by Cohort, 169
2-7 Estimated Female-to-Male Ratio of Publication, 172
2-8 Female-to-Male Odds Ratio of Post-Baccalaureate Career Paths by Family Status, 173
2-9 Comparison between Conventional Thinking and Our Findings, 174

Boxes

1-1 Meta-Analysis, 12
1-2 Stereotype Threat, 33
1-3 The Economist's Perspective, 57
1-4 Bias Avoidance Behaviors, 62
1-5 Pioneers Have Predictable Problems, 65

Introduction

The convocation was organized in four main sessions: biological components of success in science and engineering, social components of success, institutional structures that affect recruitment and retention of women scientists and engineers, and a final session on current institutional transformation efforts. Several major themes emerged from these sessions.

SEX DIFFERENCES IN COGNITIVE ABILITIES

The members of the first panel, "Cognitive and Biological Contributions," presented evidence of differences between males and females in the trajectories of brain development and in average performance on verbal, mathematical and spatial cognitive tasks.

Janet Hyde explained that the largest differences were seen in the extremes of performance distributions, with more males found in the top and bottom tails; even so, within-gender differences were much larger than between-gender differences. She presented meta-analyses that suggested that the current stereotypes in which boys and men are believed superior in mathematics skills and girls in verbal skills should be replaced with a "gender similarities" hypothesis, in which women and men are more psychologically similar than they are different. Hyde also provided data that showed girls in Taiwan and China outperformed US boys in mathematics; at issue, she emphasized, is not whether US boys do better than US girls but that US children in general are underperforming relative to nations the United States is competing with economically.

Cognitive differences showed strong dependence on age and experience, explained Jay Giedd. He provided evidence from longitudinal MRI studies of adolescents. Bruce McEwen provided evidence of differential responses to stress and suggested that there may be sex differences in learning strategies. On the basis of data showing that the influence of experience on brain development is strong and lifelong, Giedd and McEwen independently suggested that any difference between males and females be viewed more as an opportunity for education and research than as an intrinsic constraint on cognitive capability.

There was some discussion among the panelists on whether sex differences in cognition were large enough to account for the size and nature of the discrepancies between male and female representation among academic scientists; disagreement centered on the degree to which small differences could accumulate over time and have a substantial effect on careers. In that context, Diane Halpern proposed a biopsychosocial model of development, in which experience alters the biological underpinnings of behavior, which in turn influences the types of experiences to which we are exposed.

GENDER STEREOTYPES AND ACADEMIC PERFORMANCE

Science and engineering are widely stereotyped as male domains in American culture. That context influences the performance of those not expected to succeed, explained Toni Schmader. A person's belief that he or she belongs to a group stereotyped as inferior in a given ability may, when combined with certain contextual cues, trigger a phenomenon termed *stereotype threat* by Claude Steele. When this happens, the person's cognitive performance, particularly on tests of mathematics ability among women and tests of general intellectual ability among members of racial and ethnic minorities, is negatively affected. Schmader explained that contextual factors, such as predominant stereotypes, can discourage people, especially women and minority-group members, from aspiring to and pursuing science and engineering education and careers and from taking leadership roles. They also reduce their chances of being accepted into educational programs whose admission requirements emphasize test scores.

UNEXAMINED BIAS

Pervasive unexamined bias against women in science and engineering influences evaluations of women scientists' motivation, determination, promise, seriousness, and productivity and can undermine the perception of the quality of their work throughout their careers, explained Mahzarin Rustum Banaji. Small differences in advantage can accumulate over the span of a career into large differences in status and prestige. That results in male scientists often receiving greater rewards for their accomplishments than female or minority-group scientists, said Donna Ginther.

Modern gender bias, in addition to being pervasive, is automatic, ambiguous, and ambivalent, said Susan Fiske, who presented data showing the female gender role underlying this bias is both prescriptive and descriptive, demanding from women such traits as subservience and caring, thereby limiting their ability to be perceived as effective in traditionally male roles.

CAREER TRAJECTORIES IN SCIENCE AND ENGINEERING

Several panelists—including Ginther, Yu Xie, Robert Drago, Joan Williams, and Angelica Stacy—examined the factors that affect career trajectories in science and engineering.

Since the 1970s, there has been tremendous growth in the overall number of bachelor's and doctorate degrees awarded to women, but Ginther showed that women's representation is dependent upon field. In the physical sciences and engineering, women earn no more than 20% of the doctorate degrees, while in social sciences and biology women earn no less than half of doctorates.

Independently, panelists found that the factor most detrimental to career progression was family status. Their data indicated that married women scientists are disadvantaged, particularly if they have children: they are less likely to pursue careers in science and engineering even with an advanced degree, they are less likely to be in the labor force, they are less likely to be promoted, and they are less likely to be geographically mobile.

Married men with young children are 50% more likely to enter tenure-track jobs than comparable women, said Stacy. Ginther presented data showing that regardless of field, young children significantly decrease the likelihood of women—but not men—in obtaining a tenure-track job.

BIAS AGAINST CAREGIVERS

Fiske presented data showing that American culture in general strongly stereotypes caregiving—whether of children, the elderly, or sick or disabled family members—as work appropriate to females. As described above, several panelists independently identified motherhood as the factor most likely to keep a woman with science training from pursuing or advancing in a scientific career.

Scientists are generally well aware of the bias against caregiving, and those seeking fast-track academic careers use a number of strategies to avoid damage to their careers by caregiving responsibilities, said Robert Drago. Bias avoidance disproportionately affects those who shoulder primary caregiving responsibilities. What Drago termed *productive bias avoidance* involves minimizing family commitments that interfere with career progress. The most obvious methods are to avoid marriage or delay having children. What Drago termed *unproductive bias avoidance* involves efforts to deflect attention from the family responsibilities that a person in fact carries. For example, faculty members may decline opportunities to reduce their workload or to take parental leave in order to appear dedicated to their careers.

Joan Williams described *family responsibilities discrimination.* Also known as the "maternal wall", this model aims to describe in concrete terms the unrecognized patterns of stereotyping that negatively affect women in academe, to train people to recognize this bias for what it is, and to highlight an important new trend in federal employment lawsuits of which employers must be mindful. Williams discussed the federal employment laws under which employees can sue—and employers can be sued—for family responsibilities discrimination, including Title VII of the Civil Rights Act of 1964, the Pregnancy Discrimination Act, and the Family and Medical Leave Act. In sum, Williams argued for the need to create a new model for spurring institutional change that specifically names and identifies unexamined bias and considers the risk of family responsibilities discrimination lawsuits.

INSTITUTIONAL POLICIES

The traditional tactics of increasing female representation on faculties, what Joanne Martin called *add women and stir*, do not overcome systemic issues that limit women's opportunities. Policies and practices that appear to apply equally to everyone often have very different effects on men and women because of the differences in their overall social situations.

Martin explained that the widely used 7- to 10-year tenure clock and the requirement that candidates for tenure show early promise, although ostensibly gender-neutral, often create a severe conflict with the biological clock that limits women's reproductive years, forcing women to choose between taking time out for pregnancy, childbirth, and child care and pursuing fast-track careers. Other requirements of career success such as travel, relocation, and long workdays are much more difficult for people who have major caregiving responsibilities—overwhelmingly women—than for people who do not—usually men.

ETHNIC AND RACIAL MINORITIES

Women's interest in obtaining science education and pursuing scientific careers varies among ethnic groups, explained Joan Reede. While minority-group women are more likely than white women to major in and earn a PhD in science and engineering, they are rare in academic science; and once their careers have begun, they often face dual negative stereotyping. Because of their extremely small numbers on science faculties, they suffer in an exaggerated way from the problems of isolation, high visibility, unreliable feedback, inauthenticity, lack of role models, and difficulty in obtaining mentoring and camaraderie that afflict many female academic scientists generally. In addition, as members of groups underrepresented on campus, women academic scientists are under great pressure to serve on large numbers of committees.

PIONEERS

Like their minority-group women colleagues, pioneering female faculty members—those who are among the first to be hired into a department or who are the only women in their departments—face social isolation and extreme visibility, explained Joanne Martin. They are often viewed as tokens rather than genuine colleagues. Their singular status often results in unreliable feedback and difficulty in being accepted as team members and leaders. The result is that women quit, Martin said, and minority-group women quit more often.

SUMMARY AND SUGGESTIONS

From the panelist presentations and ensuing discussion, several themes emerged, primary among them that male and female careers in science and engineering generally follow different trajectories. Panelists presented data showing that sex differences in cognitive and intellectual abilities do not account for the numerical discrepancies between women and men in faculty positions. Women and minorities, because of their small numbers in faculty and leadership positions, lack the requirements of career success including mentors, camaraderie, networking possibilities, and social support. In addition, pervasive explicit and implicit gender bias—practiced by both men and women, white and minority group members—has played a major role in limiting women's opportunities and careers. Panelists provided demonstrations and data to show that bias is a complex phenomenon that requires multiple remedies, the first among them an explicit examination of the effects of bias on evaluation. Ostensibly gender-neutral institutional policies often disadvantage women scientists, particularly those targeted at women to accommodate family caregiving responsibilities, because women who take advantage of such programs are seen as less serious than their male colleagues. Women scientists who belong to ethnic and racial minorities face additional issues of stereotyping, isolation, and tokenism.

Panelists proposed a wide range of steps that institutions can take to reduce bias and inequity against women and improve opportunities for them to succeed in academic science careers. On the whole, presentations focused in on how restructuring institutional policies could alleviate problems caused by gender bias, isolation, and caregiving responsibilities. Among the steps proposed were:

- Using new metaphors and descriptions to discuss bias, in particular calling bias or stereotyping *unexamined* places the responsibility on the person who holds or acts on the bias or stereotype.
- Training people to see and identify unexamined bias in their own and others' actions.
- Establishing flexible-time policies such as family leave, flex time, part-time tenure, and temporary stoppage of the tenure-clock; and, just as

importantly, an atmosphere that allows faculty members to take advantage of these policies without fearing damage to their careers.
- Restructuring hiring and promotion procedures to reduce bias and encourage diversity, particularly the training of search committees, deans, and department chairs to recognize and reduce bias in hiring, evaluation and promotion.
- Establishing programs to provide mentoring and support to women and other underrepresented groups.
- Changing the context of test-taking to eliminate stereotype threat.
- Continued or enhanced funding of research into social and institutional structures and field testing of methods to reduce bias and stereotype threat.

A complete summary of the presentations, including figures and references, is presented in the next section. That is followed by the papers of several of the convocation speakers and the abstracts of the research posters presented at the meeting.

NEXT STEPS

In addition to this workshop report, based on the information presented at the Convocation and other research that the study committee gathers, the committee will issue a consensus report presenting conclusions and recommendations (see http://www.nap.edu/catalog/11741.html). Several convocation participants emphasized that greater workforce diversity will strengthen the American scientific enterprise and that universities and other institutions can do much to improve the opportunities for female and minority scientists to succeed in academic science.

Section 1

Summaries of Convocation Sessions

PANEL 1: Cognitive and Biological Contributions

PANEL 2: Social Contributions

PANEL 3: Organizational Structures

PANEL 4: Implementing Policies

Each session summary consists of an abstract of the panel and edited third-person transcripts of the speaker comments. The summaries present the views and opinions of the panelists and might not reflect the views of the committee or the National Academies. Slides presented by the panelists may be found on the convocation Web site, http://www7.nationalacademies.org/womeninacademe/Convocation.html.

PANEL 1
COGNITIVE AND BIOLOGICAL CONTRIBUTIONS

Panel Summary

Gender Differences and Similarities in Abilities
Janet Hyde, Department of Psychology, University of Wisconsin at Madison

Sexual Dimorphism in the Developing Brain
Jay Giedd, National Institute of Mental Health, National Institutes of Health

**Environment-Genetic Interactions in the Adult Brain:
Effects of Stress on Learning**
Bruce McEwen, The Rockefeller University

Biopsychosocial Contributions to Cognitive Performance
Diane F. Halpern, Berger Institute for Work, Family, and Children, Claremont McKenna College

Selections from the Question and Answer Session
*Moderated by committee member **Ana Marie Cauce***

PANEL SUMMARY

The panel considered whether there are differences between males and females in brain development and in average performance on cognitive tasks and whether those differences account for the large discrepancies in male and female representation among academic scientists.

Janet Hyde, of the University of Wisconsin-Madison, proposed the "novel concept of gender similarities" in cognitive abilities, noting that the mathematical, verbal and spatial skills involved in scientific work are all gender-stereotyped. Meta-analyses of 100 studies of math ability involving 3 million persons, including nine state assessments, show that the highly touted and widely reported gender differences in mathematical ability are in fact small or insignificant.

Diane Halpern, of Claremont McKenna College, observed that men and women are in fact both similar and different and "what you see depends on where you look." The differences or similarities found depend on which tests and measures are used. She also emphasized that nature and nurture form a "false dichotomy," are not independent variables, and "do not just interact." The factors are instead "inextricably intertwined" because experience alters the biological underpinnings of behavior, and the resultant biology influences the types of experiences people have. Instead of the old two-part paradigm, she proposed a biopsychosocial conceptualization of the issue and the recognition that even small differences may have large effects over time because small effects accumulate into large ones.

Jay Giedd, of the National Institute of Mental Health, presented data from magnetic resonance imaging (MRI) studies of brain structure and development during adolescence showing both gender differences in the trajectory of brain development and the strong and lifelong influence of experience on the brain. MRI studies show "gray boxes," not individual neurons, and behavioral interpretations are therefore "speculative." The sex hormones estrogen and testosterone are present both in males and females, and play a role in brain development, although hormones are not sole factors driving sex differences in the brain. Male brains show more morphological variance than female brains, but observations are based on group averages and not individuals, and overall, the brains of males and females are more alike than different.

Panelist Bruce McEwen, of Rockefeller University, presented evidence of complex sex differences in nonhuman brain response to stress and of the brain's high adaptability and plasticity throughout the lifespan. Males and female humans differ in the processes and priorities they use in processing information. Genes, hormones, and experience exert different influences on human males and females, he concluded, but the cognitive differences between men and women appear to involve differing strategies of information processing rather than different "abilities."

GENDER DIFFERENCES AND SIMILARITIES IN ABILITIES

Janet Hyde
Department of Psychology, University of Wisconsin-Madison

Janet Hyde's presentation emphasized what she called "the novel concept of gender similarities" and focused on mathematical, verbal, and spatial abilities as basic to science ability. Those abilities are "gender stereotyped," with boys believed to excel on mathematics and spatial tests and girls on verbal measures.[1]

Hyde described the power of meta-analysis (see Box 1-1) and discussed a particularly large study of gender differences in mathematics performance that pooled the results of 100 studies that tested more than 3 million people and included a wide variety of data sources, such as assessments from nine states. Averaged over all samples of the general population, the d was equal to minus 0.05, "a tiny gender difference." Another team of investigators obtained very similar results using somewhat different meta-analytic techniques.[2]

Might there be an increasing gender gap in performance with age? Second, do the mathematics tests tap lower level math computation, or a deeper conceptual understanding of mathematics and complex problem solving, which is needed to do science?

Using meta-analytic methods to investigate these questions, Hyde found that girls are better than boys at computation by a small amount in elementary and middle school. For the deeper understanding of mathematical concepts, she found no gender difference at any age level. Finally, at the highest cognitive level, complex problem-solving, she found no gender difference in elementary school or middle school, but a small difference among high school and college students. Although that difference deserves attention, it is not large.

> *The important point is that within-gender differences are enormous compared to between-gender differences.*
> —*Janet Hyde*

One explanation for the gender difference in problem-solving favoring high-school and college-age males is the difference in patterns of course taking. Girls have been less likely to take optional advanced mathematics classes in high school, although this gender gap has closed in the last five years. Girls now take calculus in high school at the same rate as boys. Nonetheless, they are less likely to take science courses in high school than boys, especially in chemistry and physics. This handicaps girls in pursuing science careers, and it also handicaps

---

[1] For more details, figures, and references, see Janet Hyde's paper in Section 2.
[2] LV Hedges and A Nowell (1995). Sex differences in mental test scores, variability, and numbers of high-scoring individuals. *Science* 270:364-365.

> **BOX 1-1**
> **Meta-Analysis**
>
> Hundreds of studies examine gender differences in performance. Rather than conduct an additional study, one can synthesize the existing studies to find an overall outcome.
>
> *Meta-analysis* refers simply to the application of quantitative or statistical methods to combine evidence from numerous studies. Meta-analysis can tell us, when we aggregate over all the available studies, whether there really is a gender difference in mathematical ability. It can tell us the direction of the difference: do males score higher on average or do females? And it can also tell us the magnitude of any gender difference.
>
> The d statistic, or effect size, is used to measure the gender difference. To obtain d, the mean score of females is subtracted from the mean score of males in a particular study, and the result is divided by the pooled within-gender standard deviation. Essentially, d tells us how far apart the means for males and females are in standardized units. d can have positive or negative values. A positive value means that males score higher, and a negative value means that females score higher. To give a tangible example, the gender difference in throwing distance is + 1.98.
>
> In a meta-analysis, d is computed for each study, and then ds are averaged across all studies. Because meta-analysis aggregates over numerous studies, a meta-analysis typically represents the testing of tens of thousands, sometimes even millions of participants. Thus, the results should be far more reliable than those from any individual study.
>
> How do we know when a d, an effect size, is small or large? The statistician Jacob Cohen provided the guideline that a d of 0.20 is small, 0.50 is moderate, and 0.80 is large.

their performance on standardized mathematics tests, because students experience mathematical problem-solving in physics and chemistry classes.

Concerning gender differences in verbal ability, meta-analysis of 165 studies representing the testing of 1.4 million persons showed superior performance by females but the difference is very small ($d = -0.11$).[3] The question of gender differences in spatial ability, a relevant skill in many fields of science and engineering, is complicated because there are many types of spatial ability and many tests to measure them. With regard to gender differences in three dimensional

---

[3] JS Hyde and MC Linn (1988). Gender differences in verbal ability: A meta-analysis. *Psychological Bulletin* 104:53-69.

mental rotation, which is crucial in fields such a engineering,[4] two meta-analyses have been conducted. One found a large gender difference favoring males, and the other found a medium gender difference favoring males,[5] both more substantial than for mathematical and verbal abilities. That does not mean that girls cannot succeed at engineering; research shows that spatial skills can be trained.[6]

One major factor in determining mathematics performance is student high school course choice. In investigating what factors influence adolescents' choice of courses and careers, Eccles found that students value what they think they will learn in a course, and that is heavily influenced by intended career.[7] Many occupations in the U.S. are highly gender-segregated. That makes it more likely that girls will not imagine themselves in science or engineering careers and therefore they will not value mathematics or physics courses as much as boys do.

Parents play an important role. Research shows that even in elementary school, parents estimate the math ability of sons to be higher than those of daughters, despite the absence of any gender difference in actual grades or test scores at this point. One particularly impressive longitudinal study found that mothers' estimates of their 6th grader's likelihood of mathematics success predicted the child's actual mathematics career choice at age 25.[8]

Schools play a third important role on the gender difference in advanced mathematics and science performance. Research shows, for example, that hands-on laboratory experiences in the physical sciences improved the science achievement of girls but not of boys, and helped to close the gender gap in achievement.

Cultural influences at the broadest level also play a role. In a cross-national study of 5th graders' math performance,[9] one could focus on the small difference in performance between girls compared with boys. However, the bigger picture

---

[4]M Hegarty and VK Sims (1994). Individual differences in mental animation during mechanical reasoning. *Memory and Cognition* 22(4):411-430.

[5]MC Linn and AC Petersen (1985). Emergence and characterization of sex differences in spatial ability: A meta-analysis. *Child Development* 56:1479-1498; D Voyer, S Voyer, and MP Bryden (1995). Magnitude of sex differences in spatial abilities: A meta-analysis and consideration of critical variables. *Psychological Bulletin* 117:250-270.

[6]Hyde referred to a study by Sheryl Sorby and her colleagues, who have developed a multi-media software program that improves the spatial performance of students and has improved the retention of women in the engineering major from 47% to 77%. See: N Boersma, A Hamlin, and S Sorby (2005). Work in progress—Impact of a remedial 3-D visualization course on student performance and retention. *Presentation at 34th ASEE/IEEE Frontiers in Education Conference, October 20-23, 2004, Savannah, GA.* http://fie.engrng.pitt.edu/fie2004/papers/1391.pdf.

[7]JS Eccles (1994). Understanding women's educational and occupational choices: Applying the Eccles et al. model of achievement-related choices. *Psychology of Women Quarterly* 18:585-610.

[8]JE Jacobs and JS Eccles (1992). The influence of parent stereotypes on parent and child ability beliefs in three domains. *Journal of Personality and Social Psychology* 63(6):932-44.

[9]M Lummis and HW Stevenson (1990). Gender differences in beliefs and achievement: A cross-cultural study. *Developmental Psychology* 26:254-263.

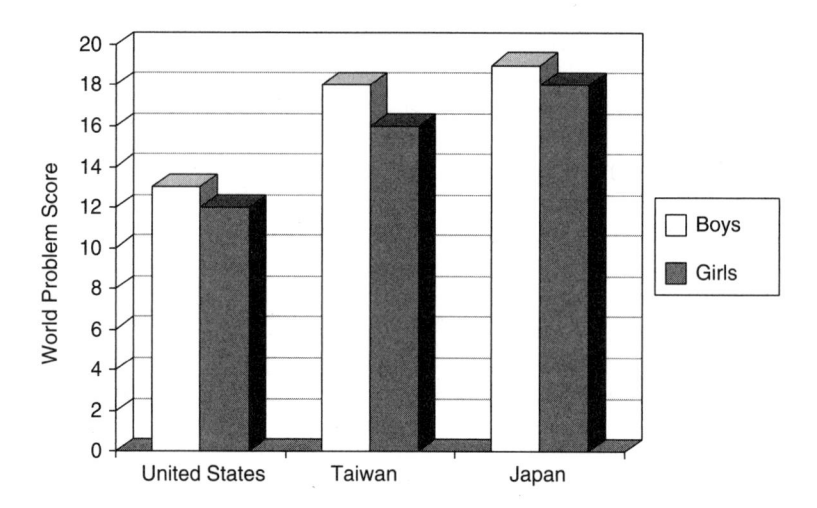

FIGURE 1-1 Cross-cultural differences in fifth-grade mathematics performance. SOURCE: M Lummis and HW Stevenson (1990). Gender differences in beliefs and achievement: A cross-cultural study. *Developmental Psychology* 26:254-263.

shows that girls in Taiwan and Japan dramatically outperform American boys. Many features probably account for this, among them differences in the way mathematics is taught, in cultural values attached to mathematics, and in different attitudes about the importance of ability vs. effort in producing excellent performance.

Another study looked at the magnitude of the gender difference in mathematics performance in different countries and correlated it with the United Nations standardized measure of gender stratification.[10] The correlation between mathematics performance and the percentage of women in the paid workforce was an impressively large –0.55 across nations. Countries with the greatest gender stratification tended to have the largest gender difference favoring males.

All those findings led Hyde to propose the gender similarities hypothesis.[11] She subjected 46 relevant meta-analyses to a meta-analysis. The studies spanned a wide range of psychological characteristics, including abilities, communication, aggression, leadership, personality and self-esteem. She found 78% of the gender differences effect sizes were small or close to 0. Psychologically, women and men are more similar than they are different. Large gender differences are found in a few cases, but the big picture is one of gender similarities.

---

[10]DP Baker and DP Jones (1993). Creating gender equality: Cross-national gender stratification and mathematical performance. *Sociology of Education* 66:91-103.

[11]JS Hyde (2005). The gender similarities hypothesis. *American Psychologist* 60:581-592.

On the basis of these data, Hyde suggested some policy recommendations: (1) a spatial learning curriculum should be instituted in primary and secondary schools, (2) colleges of engineering should have a spatial skills training program for entering students, (3) four years of math and four years of science should be required in high school or at least for university admission, (4) the mathematics curriculum in many states needs far more emphasis on real problem solving, and (5) teachers and high school guidance counselors need to be educated about the findings on gender similarities in mathematics performance, or teachers will believe the stereotypes about girls' mathematics inferiority that pervade our culture and those expectations will be conveyed to the students.

SEXUAL DIMORPHISM IN THE DEVELOPING BRAIN

Jay Giedd

National Institute of Mental Health, National Institutes of Health

Jay Giedd began by noting his focus on the adolescent brain. In child psychiatry nearly everything has different prevalences, ages of onset, and symptomatology between boys and girls and nearly every disorder is more common in boys. His studies use MRI, magnetic resonance imaging, which because it does not require radiation can be used in children to perform longitudinal studies.

> *To the MRI machine, the brain is boxes of gray or white measuring about 1 cubic millimeter. Within each of these boxes are millions of neurons and trillions of synaptic connections. Using much finer resolution microscopic techniques, one can see synapses and connections, but MRI currently cannot do that. MRI pictures and images can be quite colorful, but interpretations are necessarily speculative.*
> —*Jay Giedd*

What we call the *gray matter* consists mostly of the neuronal cell bodies, where the nucleus and the DNA are housed; the antenna-like dendrites reaching for connections to other brain cells; and the terminal branches of the axons, the location of the synapses, and the connections to other brain cells. The *white matter* is myelin, the insulation material wrapped around the axon that speeds communication between the brain cells.

Giedd and his colleagues performed longitudinal MRI scans of 2,000 subjects. They found that white matter volume increased at least through the fourth decade in women and through the third in men (Figure 1-2). At no time during development did white matter volume decrease.

The white matter has been of interest in the study of sexual brain-structure differences, or sexual dimorphism, because one of the first reports of a brain difference not related to reproduction concerned the corpus callosum, the white

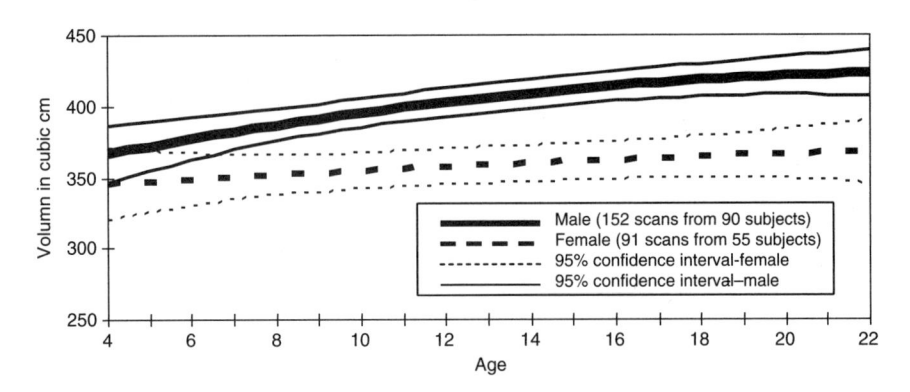

FIGURE 1-2 Longitudinal development of white matter.
SOURCE: JN Giedd, J Blumenthal, NO Jeffries, FX Castellanos, H Liu, A Zijdenbos, T Paus, AC Evans, and JL Rapaport (1999). Brain development during childhood and adolescence: A longitudinal MRI study. *Nature Neuroscience* 2(10):861-863.

matter tract connecting the two brain hemispheres. In over 100 published papers, the results are inconclusive—the corpus callosum of females is bigger than, smaller than, or not different from that of males. The key to understanding these results is in considering developmental windows. At young ages the corpus callosum is sexually dimorphic; between ages 9 and 14 it is not; and then it becomes so again. These changes happen throughout life.

> *Brain areas have intersecting developmental trajectories. This is a very important concept in how to interpret the findings. Often, the literature will combine data from people across seven or eight decades, and report that average as the difference between male and female brains.*
> —*Jay Giedd*

The most robust sex difference is total brain size. From autopsy studies, even when correcting for total body mass, male brains have been found to be about 10% larger than female brains, but bigger isn't better and size is not related to intelligence. A lot of the literature is really murky on how to account for total brain size difference.

The other part that MRI can see—the gray matter—has a distinct developmental trajectory from that of white matter. Instead of a general linear increase in volume, gray matter has an upside down "U" path in development. Changes in cortical thickness are not due to an increase in the number of neuronal cells, but to an increase in arborization, or the number of branches, twigs, and roots of existing individual neurons. Although both progressive and regressive processes occur throughout life, during childhood there is a net increase in the degree of branching and during adolescence there is a net decrease. Growth reaches a peak in the frontal part of the brain at 11 in girls and 14 years in boys. Pruning then begins: the cells and connections that are used survive and flourish, and those that are not wither and die.

There is a lot of regional variation in the process. Maturation starts in the parts of the brain needed to keep us alive, such as those controlling heart rate and breathing. The next parts of the brain to mature are those involved in processing the five senses, followed by the parts of the brain that link together the primary senses. Then there is a cascade of hierarchies linking the linkings. The final stop is the frontal lobe, which doesn't reach adult levels until about age 25.

> *By adulthood, once you correct for the total brain-size differences, the sex differences are quite subtle. But if you look at the path the brains took to get there, the differences are far more robust. It's the journey, not the destination.*
>
> —*Jay Giedd*

The most variable parts of the brain seem to be those that mature last, and are the least heritable. The structure that we have examined thus far that is the most different between males and females is the cerebellum. Because it is one of the last brain areas to mature, the cerebellum is under the influence of the environment for a long period. Accounting for overall brain size increase, the cerebellum is larger and it reaches adult volume later in males than in females. Overall, male brains have a greater variation in cortical thickness; this is a very robust phenomenon that occurs throughout the brain.

Giedd summarized with two points: First, male and female adolescent brains are much more alike than different; there is enormous overlap. Second, with regard to developmental trajectories, there are more marked sex differences. Male brain structure appears more variable. Whether the variability is biological or social in origin, the data are robust. Work is underway on the effects of sex chromosomes and hormones. In ending, Giedd emphasized that differences are group average differences, and are not to be implied as constraints for individual boys or girls.

ENVIRONMENT-GENETIC INTERACTIONS IN THE ADULT BRAIN: EFFECTS OF STRESS ON LEARNING

Bruce McEwen
The Rockefeller University

Bruce McEwen presented data on sex-based differences in the effects of stress, which have implications for learning. He and his colleagues study brain regions that are involved in memory, emotions, and executive function or decision making. He commented on the translation of animal-model studies to humans, and complemented the discussion on the continuing interaction throughout the life course between genes, hormones, and environment/experience.[12]

> *The adult brain is a very adaptable organ, and through our adult life there is a continual functional and structural remodeling.*
> —Bruce McEwen

McEwen briefly summarized the plasticity literature. When the brain is damaged, there is collateral sprouting and functional reprogramming in many cases. Even without damage, there is continual remodeling of connections with use and disuse, which has been demonstrated for the visual system and also in the motor system in terms of practice effects, such as in playing musical instruments and doing repetitive motor tasks. There are progenitor cells and even some stem cells in the adult brain; and in the dentate gyrus of the hippocampus and the olfactory bulb there is a continuous replacement of nerve cells throughout adult life. There is remodeling of the dendrites—the tree-like structures of neurons—and of synaptic connections in animals undergoing both acute and chronic stress.

Examples from an ongoing study on the prefrontal cortex illustrate the latter point.[13] In male rats that have been repeatedly stressed, neuronal dendrites become shorter and less branched and the number of synaptic connections is reduced. The overall reduction is as much as 30%, which has functional implications. However, in the amygdala, an area of the brain that is associated with fear, with aggression and emotional responses, repeated stress of the same kind causes neurons and dendrites to grow and increase their synaptic connectivity.[14] That may explain why repeated stress causes animals to become more fearful and more aggressive.

The sex hormones testosterone and estradiol have effects throughout adult life and widespread influences throughout the brain.[15] Receptors for both sex hormones are found in most brain areas, meaning that hardly any area of the brain is not influenced by circulating sex hormones. There is also evidence of a direct effect of the X and Y chromosomes on certain aspects of brain development and differentiation.

---

[12]For an overview, see BS McEwen and EN Lasley (2005). The end of sex as we know it. *Cerebrum* 7(4):65-79.

[13]JJ Radley, AB Rocher, M Miller, WG Janssen, C Liston, BS McEwen, and JH Morrison (2005). Repeated stress induces dendritic spine loss in the rat medial prefrontal cortex. *Cerebral Cortex* 16(3):313-320.

[14]A Vyas, S Bernal, and S Chattarji (2003). Effects of chronic stress on dendritic arborization in the central and extended amygdala. *Brain Research* 965(1):290-294.

[15]BS McEwen (1999). The molecular and neuroanatomical basis for estrogen effects in the central nervous system. *Journal of Clinical Endocrinology and Metabolism* 84(6):1790-1797.

Testosterone and estradiol and receptors for them are present in both males and females. Their effects in the two sexes are subtly different, depending on developmental programming. For example, estrogen affects motor coordination, vulnerability to seizures, aspects of the premenstrual syndrome or pre-menstrual dysphoric disorder, depression, vulnerability to stroke, and the amount of damage from stroke, pain mechanisms, cognitive function, and vulnerability for dementia. Estrogen influences functions both at the level of the cell nucleus through the traditional mechanism, but also through relatively newly discovered cell-surface signaling mechanisms. Similarly broad effects are seen with testosterone and other androgens in males.

There is virtually no function that is not influenced by reproductive hormones.
—*Bruce McEwen*

These broad effects should be kept in mind when thinking about how the male and female brain, with and without circulating sex hormones, responds to stressful experiences. We know that acute stress generally enhances the learning of survival-related information. Repeated stress results in adaptive plasticity. The resulting changes in dendritic branching and synaptic connectivity in areas like the amygdala, prefrontal cortex, and hippocampus, an area of the brain involved in memory, are largely reversible: when the stress ends, these effects disappear.

Recent evidence indicates that a single episode of traumatic stress results in a delayed and relatively prolonged increase in anxiety in the animal and actual growth of new synaptic connections in the amygdala and the prefrontal cortex. There is also evidence that repeated stress increases vulnerability to other traumas such as a stroke or a seizure.

In the response to stress, there are sex differences in brain remodeling. Female rats do not show the increased dendritic branching seen in the hippocampus of male rats. In contrast, dendritic branching in the amygdala appears to be enhanced by estrogen. When circulating estrogen in female rats is depleted by removing the ovaries, the stress response becomes similar to that in a male rat. Other studies have shown a greater initial effect of acute stress in the male on food intake and fear. It also appears that it takes longer for the female rat than the male rat to recover to baseline levels from a stressor.

Sex differences are neurobiologically and psychologically more complicated than we had thought. There are opposite effects in males and females of an acute stress on the conditioning of a classical Pavlovian response. Work of Gwendolyn Wood and Tracey Shors[16] shows that conditioning in male rats is enhanced by

---

[16]GE Wood and TJ Shors (1998). Stress facilitates classical conditioning in males but impairs classical conditioning in females through activational effects of ovarian hormones. *Proceedings of the National Academies of Sciences* 95(7):4066-4071.

stress. Exactly the same stress regimen in female rats profoundly suppresses conditioning. These results can be reversed by manipulating hormonal sex early in development. More recently, Shors has shown that giving the male and female rat control over the amount of shock makes the sex differences disappear.

How might some of this translate from animals to humans? McEwen suggested the key may lie in behavioral strategy. Research on rats in a water maze, where they have to swim and find a hidden platform to rest on, shows that the male and females tend to use different exploratory strategies. Without spatial cues, male rats reach the platform faster. When spatial cues are provided, females decrease the time it takes to reach the platform and do as well as or better than males. Karyn Frick and colleagues put student volunteers into an outdoor spatial maze tested memory of local contextual cues.[17] Men and women did not differ in their performance in the spatial maze but women had a better memory of objects and their location than men did.

Arguments go back and forth, and the data makes it much more complicated to reach some simple generalizations.
—Bruce McEwen

In summary, McEwen explained there are sex differences that are products of genes, of hormones, and of experience throughout the life span. Males and females do respond differently to stressors, although the differences are complex and depend on the kind of stressor and the circumstances. There appears to be modulation by circulating sex hormones, at least in the animal models. What is described in the animal literature, and also perhaps in some of the human literature, is that there are differences in processing—maybe in priorities and strategies—that are far more important than what are commonly called "abilities."

BIOPSYCHOSOCIAL CONTRIBUTIONS TO COGNITIVE PERFORMANCE

Diane F. Halpern
Berger Institute for Work, Family, and Children,
Claremont McKenna College

Diane F. Halpern began her presentation referring to a paper she had written several years ago, entitled, "What You See Depends on Where You Look."[18] Whether male and female cognitive abilities seem similar or different depends on

---

[17] LJ Levy, RS Astur, and KM Frick (2005). Men and women differ in object memory, but not performance of a virtual radial maze. *Behavioral Neuroscience* 119(4):853-862.

[18] DF Halpern (1989). The disappearance of cognitive gender differences: What you see depends on where you look. *American Psychologist* 44:1156-1158.

which data are used. To address the question, whether fewer women than men have the ability to become scientists and engineers, requires an examination of how men and women are similar and different.[19]

> *We are not talking about whether men and women are similar or different, which is debatable, because in fact women and men are both similar and different. The real question is in what ways are men and women similar and different, and how to understand the relevance of the similarities and differences.*
> *—Diane F. Halpern*

Women are graduating in very high numbers with degrees in science fields, so women obviously have the innate ability to do science. But women are not graduating in equal numbers from all of the sciences. To explain this discrepancy, some people have said that women prefer biological sciences, whereas men prefer physical sciences. Alternatively, psychologists have said that women seem to prefer people-oriented careers and men prefer thing-oriented careers. Career choice and trajectory involves a complex of traits, including abilities, interests, personality variables, opportunities, and the knowledge of available career options.

Society has many sex differences. One is the wage gap, which is not just between men and women. Overwhelmingly women are poorer than men, but the largest wage differences are between women who have children and other people. Women have fewer leadership positions overall, not just in science, not just in academia, but in corporations. College students tell us gender differences are a thing of the past, but men in college spend several more hours a week playing video games than women, among many other differences.

> *We don't like to talk about sex differences. Sex differences are simply not popular. It's much more popular to talk about similarities, there is no doubt about that. But when we talk about differences, then at least we much prefer to acknowledge that they are embedded in environment. But this concept is embedded in the false idea that nature and nurture form a dichotomy. There is not a number out there that we can pin on nature or nurture. We have got to get away from the idea of a nature/nurture dichotomy and interaction, because nature and nurture are not independent variables, and they do not merely interact. We need to replace that whole idea with a model that is biopsychosocial. Nature and nurture are inextricably intertwined; they cannot be separated.*
> *—Diane F. Halpern*

---

[19]For more details, figures, and references, see Diane Halpern's paper in Section 2.

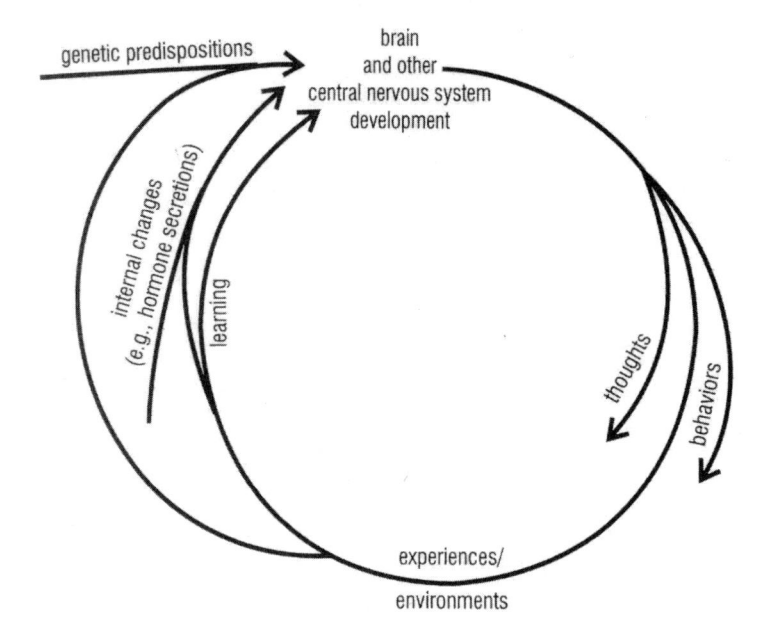

FIGURE 1-3 Biopsychosocial model.
SOURCE: DF Halpern (2000). *Sex Differences in Cognitive Abilities.* 3rd Ed. Mahwah, NJ: Erlbaum.

Experience alters the biological underpinnings of behavior which in turn influences the experiences to which we are exposed. A graphic model of biopsychosocial interactions is presented in Figure 1.3.

Some cognitive tasks show sex differences. Some of these differences are lost in aggregated data. Halpern disagreed with Janet Hyde regarding assigning values to small and large effect sizes, stating that small differences in fact accumulate to make very large differences.[20]

Some differences that favor females:
- Rapid access to and use of phonological, semantic, episodic information and long-term memory.
- Production and comprehension of complex prose.
- School grades and tests closely aligned to school curricula.
- Fine motor tasks and speech articulation.
- Perceptual threshold tasks.

---

[20]V Valian (1999). *Why So Slow: The Advancement of Women.* Cambridge, MA: MIT Press.

Some differences that favor males:
- Visual transformations and visuospatial working memory.
- Moving objects and aiming at targets.
- Fluid reasoning tasks
- Novel tasks unrelated to things that are taught in school.

Males are overrepresented in both extremes of performance—among the retarded and the gifted. That finding has been used to explain why there are fewer females in science and mathematics, but does not explain why there are fewer females in these professions overall. Not just are there fewer gifted women in science and mathematics, there are just fewer women.

International data also show sex differences. The PERLS reading study shows statistically significant effects on reading literacy at age 15, favoring girls. The mathematics test score difference is rather unimpressive and tends to be insignificant. The science test score difference at 8th grade tends to favor males and gets larger in college and graduate school as the student samples become more selective.

A test-grade disparity is part of the puzzle. Girls get higher grades in school in every subject, even when they are getting lower grades on the achievement tests. Women are graduating at a substantially higher rate than males from college, 133 women for every 100 men.

Despite those successes, women score significantly lower on many tests of science and mathematics, particularly on tests that have questions not closely related to materials taught in school. This discrepancy leads some to ask whether teachers in schools are biased against boys or whether achievement tests are biased against girls.

Cognitive processes are involved. As Bruce McEwen discussed, some have suggested that males and females are using different problem-solving strategies. Like Janet Hyde, Halpern called for education in visuospatial skills. But in trying to answer the underlying question, are there too few women with the highest levels of ability to be scientists and engineers, Halpern pointed beyond cognitive processes to a larger framework in academe: the tenure system. Marriage and having children have an adverse effect on the research productivity of women in academia.

That tenure clocks and biological clocks run in the same time zone is the more likely and proximal cause for some of these problems than cognitive differences.
—*Diane F. Halpern*

The take-home message: females and males are similar and different, depending on what is measured. The types and sizes of cognitive differences vary between men and women. Some of the measures favor females, some favor males.

There are consistent differences internationally. Halpern called for a biopsychosocial model to replace the nature/nurture dichotomy and for consideration of the larger academic and societal context.

SELECTIONS FROM THE QUESTION AND ANSWER SESSION

DR. AGOGINO: Hi, I'm Alice Agogino from the University of California at Berkeley. I have a question about how authentic these assessments or these features are in terms of actual practice and success. Janet, you mentioned the Linn Peterson study, a meta-analysis on spatial reasoning and found the greatest differences were for three-dimensional rotation, as measured by the Shepherd Test. I worked with Marcia Linn when I taught a Mechanical Engineering freshman design class where spatial reasoning skills were important. We looked at expert spatial reasoners in industry and found that they did even worse on some of those tests than the students at the lowest end of the scale. The big difference was timing. If we added 30 seconds onto a test, we got rid of a lot of the differences. We did a two hour workshop and developed strategies that improved the performance of both men and women and got rid of all the gender differences in performance on these tests. My question is, before we start creating courses, do they really matter in terms of success, and their authenticity for success in practice?

DR. HALPERN: People often ask that question. Spatial reasoning is correlated with grades in engineering schools; it's been used in dental schools as a grade predictor; and the ability to see things from multiple angles is used in imagining what a molecule will look like if you rotate it in space. In some of my own work recently we have found that males were imaging a lot of the material when they were reading it, and some of the females also. While we are teaching people how to read, we're teaching people how to do math. Cognitively this is another one of those dimensions that we have just not paid attention to in the educational system.

DR. BICKLE: Janet Bickle, formerly of the Association of American Medical Colleges, and now a career development coach. I wonder if anyone else noticed this week, a very small article in the Post that was a study of monkeys, finding that male monkeys were more likely to play with cars, and the female monkeys were more likely to play with dolls, including looking at the dolls' bottoms. And the males actually playing with the cars the way little boys do. I was wondering what sense the panelists could help us make of this type of finding.

DR. HYDE: I think partly because I'm a meta-analyst, I'm very keenly aware of how many behavioral studies in psychology don't replicate. And so, I would really want to see that study replicated before I made any interpretations, because studies like that are so quickly picked up by the media. Everybody loves them. And then there are 10 failures to replicate, and they never get attention. I think we really have to ask for the standard of replicability in a lot of these phenomena.

DR. McEWEN: I might add that while I have no comment about that particular study, it's well established in animal behavior studies on both rodents and on rhesus monkeys that there is an androgen-dependent rough and tumble play behavior which is very typical of the male of both species, and can be influenced by testosterone, and can be produced by exposure of females at the right time of development to testosterone. So, there is a phenomenon there. How it has to do with playing with any particular toy, I have no idea.

DR. GIEDD: If the studies are done well, it is a great insight into the role of socialization and media exposure and all these other sort of things, and the biology itself. So, I think it's a very worthwhile direction to pursue, if it's done well.

DR. WEYUKER: I'm Elaine Weyuker. I'm at AT&T Labs, and I'm a member of the committee. In terms of the swimming rats, one of the things I was struck by was the female rats' strategy was to swim along the edge, whereas the male strategy was to go down the middle and to look. But one of the other things I noticed was that you stuck the platform in the middle. And so, had you stuck the platform at the edge, it sounds to me like the female rats would have been the stars. Are we using as measures of "success" the things that the women don't do as well?

DR. McEWEN: The point you make gets back to this idea of strategy, and obviously, the way you set up the task can give you different results. I can give you more kinds of experiments not involving that swimming task, where again, you can establish that there are not only sex differences, but also giving estrogen to ovariectomized female rats actually improves their choice of a place strategy over a response strategy, perhaps by enhancing the function of the hippocampus over the function of other brain areas.

DR. WEYUKER: And what the measure of success is.

DR. McEWEN: Yes, that's a good point. But like the example from Karyn Frick's studies, when you are looking at the memory of location and identity of objects, on the average the women did better than the men in remembering these things. That may contribute to the success of women in handling certain kinds of spatially-related, contextual tasks where they have to remember locations of things in order to make choices.

DR. GARMIRE: I'm Elsa Garmire from Dartmouth College. The subject of this convocation is women in academe. From my point of view, I would imagine that most of women in academe would be in perhaps let's say the top 20% of whatever group that you are investigating. And what I want to know particularly in the meta-analyses, which seem to give you the average of all humans, have there been studies that have looked at the top 20%, and compared the top 20% of males to the top 20% of females in any of these studies?

DR. HYDE: There is a series of studies originally begun at Johns Hopkins by Julian Stanley and Camilla Benbow of gifted youth. They recruit mathematically precocious children in the seventh or eighth grade who score 700 or more on the mathematics SAT. Stanley and Benbow do find a disproportionate number of boys in their group compared with girls. I have never been able to pin down exactly how they recruit them though, because for example, if it's partly by teacher recommendation, then you wonder if teachers don't tend to see more mathematical talent in boys, even when it's present in girls as well.

DR. GARMIRE: Yes, but they have started out already selecting. What I'm suggesting is in all of these studies, if one went back and said, okay we are not going to look at the average for everyone, we're going to fit everyone to a bell curve, and then take the top 20% of that data, I think you could do a meta-analysis without any pre-selection of people and analyze exactly how males and females compare in the upper reaches.

DR. CAUCE: Those are a series of studies that I'm fairly familiar with. Part of what is interesting is that there are many more men in both tails of the performance distribution. But what is interesting is that even though you have more men than women in the tails, if you look at the differences in the career trajectories of the men and women in the upper tail, so we are talking about the upper 1% in terms of mathematics talent and ability, a much higher percentage of those men follow the trajectory into mathematics and science. Women are much more likely to go into particularly medicine and law than in science. I'm not aware of any studies that have tried to particularly truncate at about 20 or 25%.

DR. VOGT: Hi, Christina Vogt, National Academy of Engineering. I think that we need to look a little bit more at social determinants of engineering and science careers than spatio-visualization skills.

DR. HYDE: I agree, and I think some of the panels later today are going to be getting at some of the factors like that, so it's definitely important.

DR. CAUCE: I couldn't agree with you more. There is no question but that workplaces and how people react to them are different. But then also there is some work that suggests there might be some biologically based differences in motivations, so, that women would be motivated more towards going into social careers, which are defined, and I would say erroneously, as being non-science careers.

DR. SAENZ: My name is Delia Saenz. I'm a social psychologist at Arizona State University, and I do work on tokenism. Much of that work, at least in the early part of my career, demonstrated that tokens suffer cognitive deficits. I remember when I found the first result, I wanted to hide the research, because I thought, okay, nobody is going to want to hire women or minorities, because they will bring them in, and they will do poorly, not because of their capacity, but rather because of the environmental configuration that is having them concerned with self-presentation. One of my best friends said, you know what? You got the same finding for males and for white males. So, it's not a matter of who you are, but the context.

I agree with what you all suggested earlier, that if you match the person to the task, and you have a good fit, things will go better. And in fact, my more recent work on tokenism suggests that there are cognitive surfeits if you are a token. So, because you are concerned with self-presentation, you're better able to take perspectives, and you are good at negotiating, which is a good thing, and it happens for women and minorities, as well as for males and whites. That's very exciting. So, we will get to the point where we are not just focusing on differences in ability, but differences in outcome, differences in being able to make a living and having your contributions validated.

PANEL 2
SOCIAL CONTRIBUTIONS

Panel Summary

Implicit and Explicit Gender Discrimination
Mahzarin Rustum Banaji, Department of Psychology,
Harvard University and Radcliffe Institute for Advanced Study

Contextual Influences on Performance
Toni Schmader, Department of Psychology, University of Arizona

Interactions Between Power and Gender
Susan Fiske, Department of Psychology, Princeton University

Social Influences on Science and Engineering Career Decisions
Yu Xie, Department of Sociology, University of Michigan

Selections from the Question and Answer Session
*Moderated by committee member **Alice Agogino***

PANEL SUMMARY

The panel examined the role of bias, discrimination, and personal preference in cognitive performance, evaluation of ability, and career preferences. Mahzarin Rustum Banaji, of Harvard University, used an audience-participation technique based on her widely utilized, computer-based data collection techniques to demonstrate the unconscious, automatic, and unintentional nature of implicit biases and their dissociation from conscious beliefs. An audience composed overwhelmingly of female scientists, scholars, and government and university administrators displayed biases widespread in the culture that assume that science and mathematics are masculine and home and family feminine. The pervasiveness of such unconscious or implicit bias is important because a meta-analysis indicates that biases predict action. Biases are nonetheless malleable, Banaji continued, with the science now providing insight into how even implicit stereotypes can be changed.

Toni Schmader, of the University of Arizona, presented data on stereotype threat, the negative effect of stereotyping on test performance. Context, such as framing a test as a measure of ability or reminding test-takers of gender, can trigger stereotype threat that lowers performance and self-confidence and can discourage women and minority-group members from seeking mathematics and science careers or leadership roles important to career success. Reducing stereotype threat can release cognitive resources needed for peak performance.

Susan Fiske, of Princeton University, explored the interaction between power and gender as revealed in modern gender bias, which is automatic, ambiguous, and ambivalent. Ambiguity reveals itself in several ways: in shifting standards; in the short-list problem, in which women are nominated but not selected for high posts (giving decision makers "moral credentials" for short listing the women even if she was not chosen); and in women's alleged "lack of fit" for posts traditionally considered male. The traditional gender role of female subservience is not only descriptive but also prescriptive, and women in the workplace who defy its limits are punished. Objective standards are needed to deal with ambiguous bias. Ambivalence reveals itself in two types of sexism— hostile and benevolent— which correlate, respectively, with the stereotype of nontraditional women as "not nice" and traditional women as "not competent"; this leads to the Catch 22 that women are either liked and not respected or respected and not liked. To overcome ambivalent bias, women must focus on gaining respect, often with costs to their rated likeability.

Explaining the discrepancy between men and women in science requires giving up the "naïve idea of finding simplistic explanations," according to Yu Xie, of the University of Michigan. The life-course approach—a perspective that recognizes interactive effects, individual variations and the cumulative nature of these effects—forces rejection of several commonly offered explanations. The "critical filter" hypothesis that inadequate high-school mathematics training handicaps women or the fact fewer women than men score in the upper percentiles in mathematics ability does not explain why fewer women than men major in science. The metaphor of the science career as a pipeline also falters because it incorrectly assumes that one can only leave science and not come back to it. The so-called "productivity puzzle," which argues that women scientists are systematically less productive than men, vanishes if contextual factors are held constant. The factor most likely to prevent a woman with science training from pursuing a scientific career is children. The discrepancy between men and women in science has deep social, cultural, and economic roots.

IMPLICIT AND EXPLICIT GENDER DISCRIMINATION

Mahzarin Rustum Banaji
Department of Psychology,
Harvard University and Radcliffe Institute for Advanced Study

Mahzarin Banaji focused her presentation on an invisible form of bias—implicit bias—and her work using the Implicit Association Test (IAT). She began by quoting a colleague who had said, "Women are not being kept out of science by force, so they must be choosing not to enter, presumably because they don't want to, presumably because by and large, they don't like these fields, or on average don't tend to excel in them, which is nearly the same thing."

Psychologists have spent careers in trying to understand the meaning of words like *choose*, *want*, and *like*. They are complicated. Much of the way we behave happens outside conscious awareness. Many of our thoughts and feelings arise in an automatic and unintentional fashion.

> *Our evolutionary history sets us up to have a particular way of looking at things. We are immersed in a larger culture that teaches us the associations between large categories and particular attributes. We need a much finer-grained understanding of what we mean by environment.*
>
> —*Mahzarin Banaji*

Banaji and her colleagues have done experiments using the IAT in which they ask people to look at pictures and say what they see.[21] Gender is not verbalized, but it affects decisions about the next object viewed. Gender is evoked quite outside conscious awareness and is associated with the image object. Banaji also has used the IAT to research the strength of the association between sex and

---

[21] For more on the Implicit Association Test, see https://implicit.harvard.edu/implicit/.

categories such as mathematics and science vs. the humanities. Both men and women show high association between self and their gender group. For men, there is a positive association between maleness and mathematics. For women, there is a negative association between femaleness and mathematics. Furthermore, stronger me-female connections are correlated with stronger female-does-not-equal-mathematics connections.[22]

There is a large difference between *perceived* or *believed* difference and *actual* difference in mathematics performance. The bias associating maleness with mathematics has a *d* of 1.5-2. The performance differences meta-analyses reported by Janet Hyde show a *d* of 0.05. These biases are large and pervasive.

A signature of implicit biases is that they contradict conscious beliefs. It is not that a person does not know that mathematics is stereotyped as male, and that home is stereotyped as female. Rather, people taking the IAT who try explicitly to associate each sex equally with each category cannot. This contradiction is of interest for a variety of reasons. Most interesting, it shows the deviation from where we want to be.

Implicit biases have predictive power. A meta-analysis of close to 100 IAT studies showed that the magnitude of the bias demonstrated in experimental conditions accurately predicts a person's behavior in nonexperimental situations.[23]

> **This kind of test tends to predict attitudes toward affirmative action. It tends to predict whether one will hire somebody who is a female or not, and so on. We need more science to show us how these kinds of associations affect our behavior.**
>
> —*Mahzarin Banaji*

The optimistic part of this message is that these biases are malleable—in ways that many of us never could have imagined. Put girls and boys into a room where there are signals that make the association between mathematics and women. The biases will change so that girls will make the association between women and mathematics within a period of a few minutes, overcoming temporarily what has been learned over a long period.

Attitudes and beliefs are malleable and easy to change if we know what to do. It may not take much effort to fix the problem once we know what to do.[24]

---

[22]BA Nosek, MR Banaji, and AG Greenwald (2002). Math = male, me = female, therefore math ≠ me. *Journal of Personality and Social Psychology* 83(1):44-59.

[23]W Hofmann, B Gawronski, T Gschwendner, H Le, and M Schmitt (2005). A meta-analysis on the correlation between the Implicit Association Test and explicit self-report measures. *Personality and Social Psychology Bulletin* 31(10):1369-1385.

[24]For example, see J Kang and MR Banaji (2006) Fair Measures: A behavioral realist revision of "affirmative action." *California Law Review* 94:1063-1118.

CONTEXTUAL INFLUENCES ON PERFORMANCE
Toni Schmader
University of Arizona

Toni Schmader followed up on pervasive, implicit biases and focused on the social context in which contending with these biases can shape women's performance on many of the types of tasks that were presented in the first panel discussion.

One of the lessons that we have learned from social psychology is that we have a tendency to look at what a person does and to assume that the main variable responsible for their behavior is them.
—Toni Schmader

In a classic study, observers watched as one participant struggled to answer esoteric trivia questions asked by another participant.[25] The observers knew that the two participants had been randomly assigned to either ask or answer questions. They also knew that the questions were unreasonably difficult, but they still had a bias toward assuming that the person answering the questions was less competent and less intelligent than the person asking them. These data make the point that we tend to want to infer people's ability from their performance even when we know that the social context stacks the deck against them. To what degree do these implicit biases and gender stereotypes that assert women's incompetence in mathematics, science, and engineering undermine women's ability to perform?

Stereotype threat applies as well to women performing on a difficult mathematics test. In one of Schmader's recent studies,[26] men and women in one condition were told that their task would yield a diagnostic measure of mathematics ability that would be used to compare men's and women's scores; in this condition, there was a gender gap similar to that seen in SAT scores shown by Diane Halpern. But in a second condition, a second group of students given the same set of word problems were told that it was just a problem-solving exercise, with no mention of a test, mathematics, or ability; here, women's performance on the test was significantly better and not different from that of their male peers regardless of whether differences in SAT were controlled for (Figure 1-4).

Results like those should make us question whether the kinds of differences we see in performance measures can be adequately accounted for by underlying

---

[25] L Ross, TM Amabile, and JL Steinmetz (1977). Social roles, social control, and biases in social perception. *Journal of Personality and Social Psychology* 35:485-494.

[26] M Johns, T Schmader, and A Martens (2005). Knowing is half the battle: Teaching stereotype threat as a means of improving women's math performance. *Psychological Science* 16:175-179.

> **BOX 1-2**
> **Stereotype Threat**
>
> In 1995, Claude Steele and Josh Aronson[a] published an influential article in which they demonstrated a phenomenon they called *stereotype threat*. Stereotype threat occurs when people feel that they might be judged in terms of a negative stereotype or that they might do something that might inadvertently confirm a stereotype about their group.
>
> When any of us find ourselves in a difficult performance situation, especially one that has time pressure involved, we might recognize that if we do poorly, others could think badly about our own individual abilities. But if you are a woman or minority-group student trying to excel in science, there is the added worry that poor performance could be taken as confirmation that group stereotypes are valid.
>
> In their first series of studies, Steele and Aronson set out to ask whether you could change a minority-group student's ability to perform on a difficult intellectual task by simply changing the context, for example, how the task is described. They had white and black college students at Stanford University come into a laboratory to complete a set of difficult questions taken from the Graduate Record Examination (GRE). Half the participants were told that the test would measure verbal ability—the same kinds of instructions that students might expect to get before taking the GRE. They found the same type of race gap in test scores that is often seen on standardized tests. For a second group of students, the same task was described as a laboratory exercise. Under these more neutral conditions—in which no reference was made to race, ability, or a test—African American students performed significantly better; their performance was not different from that of their white peers.
>
> ---
>
> [a]CM Steele and J Aronson (1995). Stereotype threat and the intellectual test performance of African Americans. *Journal of Personality and Social Psychology* 69:797-811.

differences in ability. If differences in ability explained the gender gap or the race gap, as least with these kinds of samples, it should not be so easy to erase or reduce that gap by simply changing how the test is described.

We know that contextual cues, such as how a test is described, can be one type of variable that can lead to stereotype threat. Research suggests that other types of situational cues that can lead to the same processes. For example, something that merely reminds people of their gender or race can be enough to produce

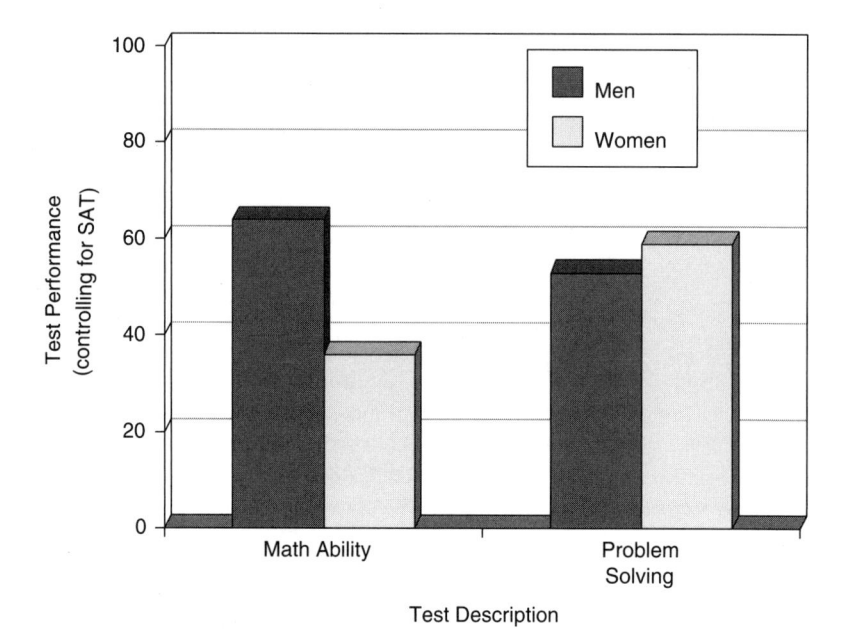

FIGURE 1-4 Gender differences in mathematics performance.
SOURCE: M Johns, T Schmader, and A Martens (2005). Knowing is half the battle: Teaching stereotype threat as a means of improving women's math performance. *Psychological Science* 16:175-179.

these processes.[27] In some experiments, simply having a woman answer a questionnaire about gender issues before taking a mathematics test leads to a significant reduction in performance.[28]

It is true that a lot of these experiments have been done with college-aged populations, but the effects have been replicated in younger age groups as early as elementary school.[29] Replications are also seen in more natural settings such as classroom environments.[30]

These data tell us that context can shape performance on test scores. But

---

[27] CM Steele and J Aronson (1995), Ibid; M Inzlicht and Ben-Zeev (2000). A threatening intellectual environment: Why women are susceptible to experience problem-solving deficits in the presence of men. *Psychological Science* 11:365-371.

[28] M Shih, TL Pittinsky, and N Ambady (1999). Stereotype susceptibility: Identity salience and shifts in quantitative performance. *Psychological Science* 10:80-83.

[29] N Ambady, M Shih, A Kim, and TL Pittinsky (2001). Stereotype susceptibility in children: Effects of identity activation on quantitative performance. *Psychological Science* 12:385-390.

[30] J Keller (2002). Blatant stereotype threat and women's performance: Self-handicapping as a strategic means to cope with obtrusive negative performance expectations. *Sex Roles* 47:193-198.

what about other types of variables? Does women's or girls' preference or interest in mathematics reveal conscious choice? Research indicates that implicit biases can shape what students believe about what they are capable of and then what they are interested in.

In a study at the University of Arizona, Schmader and colleagues asked female science majors to rate the degree to which they agreed with statements about inherent differences in abilities between men and women.[31] Most students tended to reject beliefs about inherent sex differences in abilities, but some wondered whether such innate differences might exist. In her data set, students who tended to agree with statements about inherent sex differences reported having less confidence in their own abilities in their science majors, lower self-esteem about their performance, and less interest in attending graduate school in their major field.

Where do the stereotypes come from? Even if parents and teachers are well intentioned and try to guard their students against these kinds of beliefs, children from a very early age are bombarded by messages that say what a feminine woman should be like. Recent evidence suggests that even experimental exposure to these mass media affect a woman's stated interest in pursuing a career in science or engineering. Female college students were shown television commercials that were neutral or that portrayed women as stereotypically feminine. After exposure to the stereotypic ads, women reported less interest in science and mathematics careers than in language-based careers. In a later study, after exposure to the stereotypic ads, women also reported less interest in taking on a leadership role and instead preferred a more subordinate role in which they would be taking direction from others.[32]

Together, those data suggest that the context, namely stereotypes that exist in the environment, can lead to lower test performance and maybe shape lower confidence, can lead some women to develop less interest in pursuing science- and mathematics-based careers even when they major in those fields, and maybe can shape students' interest in taking on the leadership roles that are necessary for success in academic research.

Having provided some evidence that context can shape performance, how do we go about closing the gender gap? Context can be changed through a combination of social policy designed to create threat-free environments and educational strategies to try to teach both students and mentors about the kinds of circumstances in which bias can exist. The mere presence of successful and competent

---

[31] T Schmader, M Johns, and M Barquissau (2004). The costs of accepting gender differences: The role of stereotype endorsement in women's experience in the math domain. *Sex Roles: A Journal of Research* 50:835-850.

[32] PG Davies, SJ Spencer, DM Quinn, and R Gerhardstein (2002). Consuming images: How television commercials that elicit stereotype threat can restrain women academically and professionally. *Journal of Personality and Social Psychology* 33:561-578; PG Davies, SJ Spencer, and CM Steele (2005). Clearing the air: Identity safety moderates the effects of stereotype threat on women's leadership aspirations. *Journal of Personality and Social Psychology* 88:276-287.

women in science and engineering can send a signal that women can be capable in these fields. In controlled laboratory experiments, there is a gender gap in mathematics test scores when a study is run by a competent man; when the study is run by a competent woman, that gender gap is reduced.[33]

In addition to changing the gender composition of faculty leadership positions, we can change the gender composition of the classroom. We can try to close the gap through education. We need to teach our educators to be wise mentors, to speak out against the stereotypes in front of students. Research suggests that stigmatized students are most likely to be motivated to work on their mistakes and grow from past experiences if they receive feedback that provides a combination of high standards for performance and communication from the educator that students are capable of meeting them.[34]

> *We can emphasize skill over ability and frame learning as part of the incremental process where tests measure progress towards goals. We can try to foster a sense of belonging among young women in the sciences. Often, when members of stigmatized groups face difficulty or challenges, they take it as a sign that they are in the wrong place, that they don't belong. By helping students to see that learning and diversity are natural parts of the educational process, we can help them to adjust their interpretation of the situations they encounter.*
> —Toni Schmader

A year-long intervention study tested the effectiveness of those kinds of educational messages.[35] College students mentored three groups of 7th-grade students. One group was taught by the college students over the course of the school year that intelligence is an incremental skill that grows with effort. The second group was taught that experiencing difficulties is a normal part of educational growth. And the third group, a control group, was given anti-drug messages. At the end of the school year, there was a statistically significant gender difference in mathematics test performance only among the students who were in the control group. There was no measurable gender difference in test performance in the two groups that received the educational messages.

Another way to inoculate students through education is by unveiling the effects that implicit biases and stereotype threat can have on a woman's performance and anxiety. When women are facing difficulty in a specific performance

---

[33] DM Marx and JS Roman (2002). Female role models: Protecting women's math test performance. *Personality and Social Psychology Bulletin* 28:1183-1193.

[34] GL Cohen, CM Steele, and LD Ross (1999). The mentors' dilemma: Providing critical feedback across the racial divide. *Personality and Social Psychology Bulletin* 25:1302-1318.

[35] C Good, J Aronson, and M Inzlicht (2003). Improving adolescents' standardized test performance: An intervention to reduce the effects of stereotype threat. *Applied Developmental Psychology* 24:645-662.

situation such as taking a standardized test, they may interpret that difficulty as a sign that they are not capable.

> *By being able to externalize anxiety, women might by able to free up the cognitive resources that are necessary to focus on the task at hand.*
> *—Toni Schmader*

In the earlier-described math test[36] one group of students were told that their task would yield a diagnostic measure of mathematics ability that would be used to compare men's and women's scores, and a second group of students were told that the task was just a problem-solving exercise. There was a third condition to that experiment. Students were told that the test they were taking was a diagnostic measure of mathematics ability, and that their performance would be used to compare men's and women's scores—the same conditions that led to performance decrements in the first group. However, they were also informed about stereotype threat and reminded that if they were feeling anxious while taking the test, it might be a result of external stereotypes and not a reflection of their ability to do well. Under those conditions, women's performance was significantly increased and not significantly different from that of their male peers (Figure 1-5).

We need additional research and additional funding to identify the precise mechanisms that account for the effects of stereotypes. But we also need funding to develop more field-based interventions that would put into practice some of the available ideas to test their effectiveness in closing the gap.

In closing, Schmader discussed the implications of the contextual approach for such policy issues as affirmative action. She spoke about the need to create more diverse learning environments, to make sure that there are women and minority group members both in the student body and on faculty. To the degree that affirmative-action policies can help to ensure that we have that kind of diversity of representation, they can create not only a threat-free environment for women and others who are socially stigmatized in science and engineering but also a more diverse learning experience for everyone.

> *In light of implicit biases and contextual effects on performance, affirmative action can do more. If the difference that we see in standardized test scores can be explained by contextual factors that are systematic and that affect men and women differently, it seems reasonable for admissions committees to take that into account when they evaluate student applications.*
> *—Toni Schmader*

---

[36] M Johns, T Schmader, and A Martens (2005). Knowing is half the battle: Teaching stereotype threat as a means of improving women's math performance. *Psychological Science* 16:175-179.

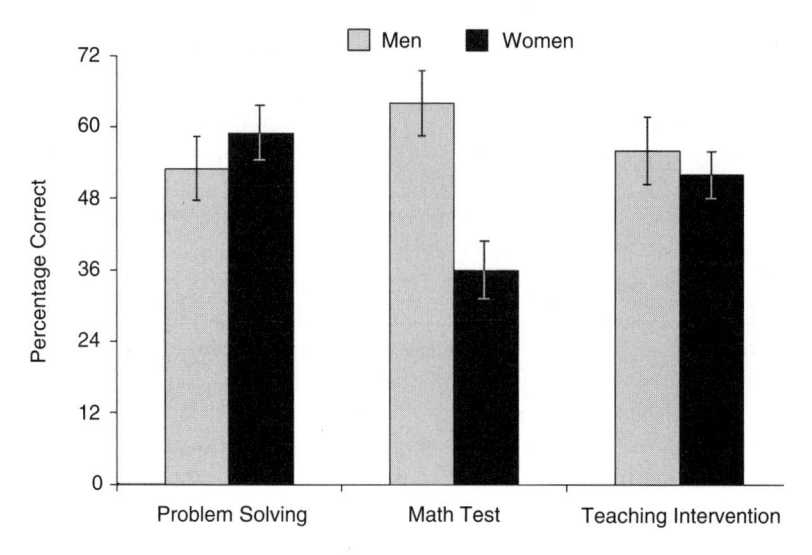

FIGURE 1-5 Teaching about stereotype threat inoculates against its effects.
SOURCE: M Johns, T Schmader, and A Martens (2005). Knowing is half the battle: Teaching stereotype threat as a means of improving women's math performance. *Psychological Science* 16:175-179.

INTERACTIONS BETWEEN POWER AND GENDER

Susan Fiske
Department of Psychology, Princeton University

Susan Fiske discussed the relationship between gender stereotyping and various manifestations of power in the context of women moving into science careers, particularly the effects of ambiguous and ambivalent biases.

Modern forms of gender bias are not your grandmother's version of gender bias.

—Susan Fiske

Several studies have shown how gender stereotypes and prejudice are ambiguous. One is from Monica Biernat's work on shifting standards.[37] Her work showed that people will say that a candidate is "really good for a woman," but

---

[37]M Biernat and ER Thompson (2002). Shifting standards and contextual variation in stereotyping. *European Review of Social Psychology* 12:103-137.

when comparing the woman to a man, find her lacking. Such judgment depends on whether standards are subjective or objective. A number of validated judgment dimensions need to be considered in the standards that people use.

Terri Vescio has demonstrated the related "short list" problem.[38] Women may be nominated and appreciated and put on short lists for opportunities, but when a choice has to be made, women are not picked. People gain moral credentials for developing unbiased short lists, but in making a final decision they weight things in favor of the status quo.

Madeline Heilman's work on lack of fit has demonstrated that if the predominant model is that managers are male or scientists are male, then women somehow don't fit if they seem "like women."[39] But, if women try to fit by acting like men, they are not liked very much, and that doesn't work either.[40]

It comes down to what Barbara Gutek has called sex-role spillover: people have implicit expectations that men are going to act like men and women act like women in the workplace. She shows this assumption can lead to sexual harassment.[41] When women are agentic—assertive and controlling—and do not act like traditional women in the workplace, there is backlash, as shown by Alice Eagly and Laurie Rudman.[42]

Gender stereotypes are not just descriptive, they are prescriptive. It's not just how women are, it's how women are supposed to be. And women who behave out of role are punished for it.

—Susan Fiske

Gender stereotyping is ambiguous. People cannot easily know when they are the objects of gender stereotyping nor, for that matter, when they are perpetuating stereotypes. It is very hard to be on a committee that is making a decision and to decide whether the decision is biased or not, because stereotyping is ambiguous. It is no longer somebody saying a woman cannot be hired. It is much more subtle. That is why the numbers are important. We have to look at the education and

---

[38] M Biernat and TK Vescio (2002). She swings, she hits, she's great, she's benched: Shifting judgment standards and behavior. *Personality and Social Psychology Bulletin* 28:66-76.

[39] M Heilman (2001). Bias in the evaluation of women leaders. Description and prescription: How gender stereotypes prevent women's ascent up the organizational ladder. *Journal of Social Issues* 57(4):657-675.

[40] Price Waterhouse v. Hopkins, 490 U.S. 228 (1989).

[41] BA Gutek and B Morasch (1982). Sex-ratios, sex-role spillover, and sexual harassment at work. *Journal of Social Issues* 38:55-74.

[42] LA Rudman and P Glick (2001). Gender effects on social influence and hireability: Prescriptive gender stereotypes and backlash toward agentic women. *Journal of Social Issues* 57(4):743-762; AH Eagly (2004). Few women at the top: How role incongruity produces prejudice and the glass ceiling. In *Identity, leadership, and power*, Eds. D van Knippenberg and MA Hogg. London: Sage Publications.

workforce pyramid: many women major in science and engineering, fewer women go to graduate school, fewer become assistant professors, still fewer become tenured professors, and even fewer become full professors and deans. Women are bailing out at every stage. The incoming cohorts, while they are making a difference, are not going to make up for this loss of talent.

What's so special about sex? The things I've been mentioning are true for other kinds of stereotypes. They are true for racial stereotypes, too, for the most part. The difference is that men and women are wonderfully and horribly, depending on the circumstances, intimately interdependent. That is the source of great joy and great personal tragedy. Men and women have personal power in their interdependence, but it's a different kind of power from the societal power that men have in general. And that leads to profound ambivalence in gender prejudice.
—Susan Fiske

Fiske and Peter Glick have developed a theory of ambivalent sexism, which is built upon the concepts of hostile and benevolent sexism. Male dominance leads to the possibility of *hostile sexism*, which is what people commonly associate with the term "sexism." Hostile sexism is targeted particularly at nontraditional women, that is, women who are perceived to challenge men and male dominance. But a different kind of sexism had not been identified before in the psychology literature. Intimate interdependence leads to *benevolent sexism*, attitudes that are experienced as favorable toward women serving in traditional roles, such as homemakers.[43]

Together, hostile sexism and benevolent sexism maintain the status quo. Both can be measured with the Ambivalent Sexism Inventory, validated with people all over the world. What Fiske and Glick find is that hostile sexism correlates with negative stereotypes of nontraditional women and benevolent sexism correlates with positive stereotypes of traditional women. Benevolent sexism predicts positive evaluations of homemakers and negative evaluations of career women. Across many countries, men score higher on hostile sexism than women do. Women do not score zero; they can be sexist, too. But on average, hostile sexism is stronger for men. With benevolent sexism, the difference is much smaller. Levels of hostile and benevolent sexism are correlated with United Nations indices of human development.[44]

---

[43]P Glick and ST Fiske (1996).The Ambivalent Sexism Inventory: Differentiating hostile and benevolent sexism. *Journal of Personality and Social Psychology* 70:491-512.

[44]United Nations Human Development Programme (1995). *Human Development Report 1995.* New York: Oxford University Press, http://hdr.undp.org/reports/global/1995/en/; P Glick, S Fiske, A Mladnik, JL Saiz, D Abrams, et al. (2000). Beyond prejudice as simple antipathy: Hostile and benevolent sexism across cultures. *Journal of Personality and Social Psychology* 79:763-775.

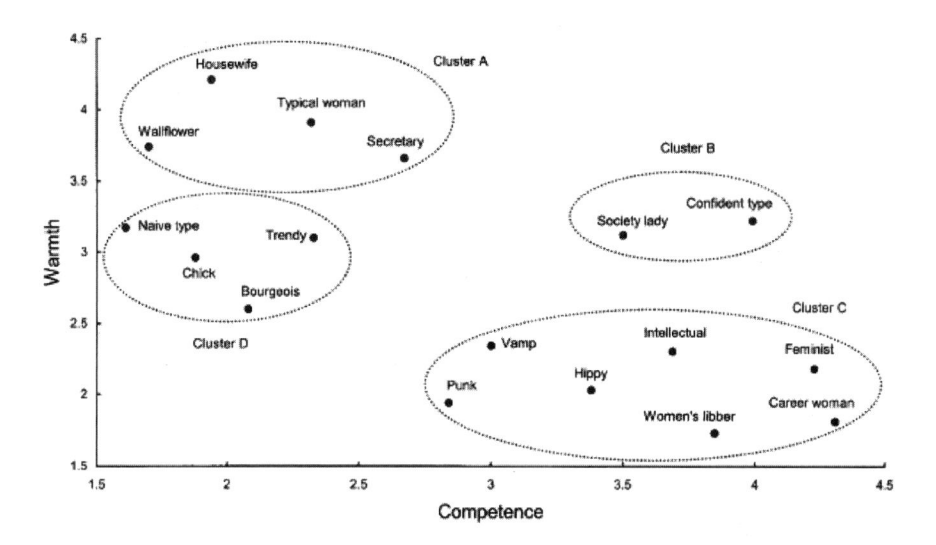

FIGURE 1-6 Fiske et al.'s Stereotype Content Model applied to subtypes of women.
SOURCE: T Eckes (2002) Paternalistic and envious gender stereotypes: Testing predictions from the stereotype content model. *Sex Roles* 47(3-4):99-114.

The tension between being liked and being respected—homemakers are liked but disrespected and career women are respected but disliked—maintains inequality by confining women's roles, as shown by Thomas Eckes.[45] Combining his work with Fiske and Glick's stereotype content model, as shown in Figure 1-6, shows housewives categorized as incompetent along with disabled people, senior citizens, and unemployed people. Career women and feminists are categorized as competent with managers, politicians, and millionaires. The liked but disrespected homemakers are protected and helped, but also excluded and neglected. With career women, who are respected but disliked, others will cooperate, associate, and go along to get along when they have to. But when the chips are down, career women are more likely than men to be attacked and sabotaged.

Fiske emphasized the need to recognize the tightrope that women are walking not to be too feminine, not to be too masculine, but somehow managing to juggle these gender tensions. With respect to ambivalence, most out-groups really do not care if you like them or not. They want to be respected, so that they can be

---

[45]T Eckes (2002). Paternalistic and envious gender stereotypes: Testing predictions from the stereotype content model. *Sex Roles* 47(3-4):99-114.

promoted. Fiske suggested establishing careful standards by which people are evaluated to mitigate the effects of automatic, ambiguous, and ambivalent gender bias.

SOCIAL INFLUENCES ON SCIENCE AND ENGINEERING CAREER DECISIONS

Yu Xie
Department of Sociology, University of Michigan

Yu Xie based his presentation on a book he researched and wrote with Kimberlee Shauman, an associate professor of sociology at the University of California, Davis.[46] He highlighted major findings from the book.[47]

Earlier studies on sex differences in career trajectories examined only subsets of scientists and engineers, such as high school students, college students, graduate students, and practicing scientists. Xie and Shauman analyzed 17 large, nationally representative datasets that spanned the career. They adopted a life-course approach, which recognizes interacting effects across multiple domains in a life, such as education, family and work. What we do in one domain of life affects what we do in other domains, so these factors are interrelated and cumulative. What happened before affects what happens now. What is happening to you now affects what will happen later. We call this path dependence.

To implement a life-course perspective on the study of gender in science, it is necessary to pay attention to data. Ideally, we would have data that span the entirety of a career from early ages to retirement. Not only do we want to have a dataset so expansive in scope, we also would like to have longitudinal data that follows the same individuals over their life course. Because scientists are only a small proportion of the labor force in the population, it is not possible to do that. To make up for the deficiency in data sources, Xie and Shauman painted a composite picture, using some data sources from (1) students in grades 7-8, (2) high school students, (3) college students, (4) graduates who attain bachelor's degrees and master's degrees in science and engineering, and (5) individuals who work in the labor force as scientists. This approach is called a synthetic cohort analysis.

The short version of the conclusion of this study is really one word: complexity. There are no simple answers.

—Yu Xie

---

[46]For more details, figures, and references, see Yu Xie's paper in Section 2.
[47]Y Xie and K Shauman (2003). *Women in Science: Career Processes and Outcomes*. Cambridge, MA: Harvard University Press.

Xie and Shauman rejected several widely held hypotheses and claims, with which the data were not consistent.

The first rejected hypothesis was the *critical filter* hypothesis, which states that women are handicapped or disadvantaged because they are not good at mathematics in high school. The gender gap in average mathematics achievement is small and has been declining, as Hyde discussed earlier. However, males are more dispersed in the high and low ends of the achievement spectrum. The representation of girls is lower than the representation of boys in the top 5% of achievement. However, gender differences in average mathematics achievement and in high level mathematics achievement do not explain gender differences in majoring or degree attainment in science.

The second hypothesis was the *pipeline paradigm*, which assumes that we can only leave science and not come back. That is not accurate. Career processes are fluid and dynamic. Entry, exit, and re-entry are all possible. Participation gaps are greatest at the transition from high school to college. A substantial percentage of males and females express the desire to become science and engineering majors, according to attitudes assessed in high school. Some will change their mind and major in nonscience fields in college, but a fraction of them obtain science degrees. The most critical juncture is the transition between high school and college. Not only do fewer high school girls expect to major in science in college, but from this point to the first year in college, fewer females are likely to realize their expectation than males. After the first year in college, there is little difference in persistence to a degree attained.

The third hypothesis was the *productivity puzzle*. Xie and Shauman looked at practicing scientists employed as faculty members in colleges and universities. A standard claim has been that women publish slightly more than half as many papers as men. Cole and Zuckerman looked at the historical trend and at everything they could find to explain this gap, and could not explain it away. Scott Long reaffirmed the conclusion. In their reanalysis, Xie and Shauman had two major findings. First, looking at research productivity over the time from the late 1960s all the way to 1993, one sees a steady increase in women's productivity relative to men's. The steady improvement in women's research productivity suggests something deeper and broader than biology alone. Second, most of the observed sex difference in research productivity even in the earlier years can be attributed to sex differences in background characteristics, employment positions and resources, and children.

The fourth and hypothesis examined was that a *family life* hampers women scientists' careers. They found that married women with children were less likely than men or other women to pursue science careers after the completion of science or engineering education; they are less likely to persist in science; they are less likely to be in the labor force or employed and they are less likely to be promoted. These women have already attained education in science, so it is not that they cannot do the work, pass the examinations, and learn the material.

Is there a family effect? We find that marriage itself does not seem to matter much. Married women are disadvantaged only if they have children.

—*Yu Xie*

If you compare single men with single women, you do not see differences in the likelihood of whether they work or not, whether they go to graduate school rather than work, whether they are in graduate school, whether they are in science, or if they work, whether they are in science or not in science. You see the biggest gender gap when women are married and have children. Married women with children are more likely to stay out of graduate school and work. If they go to graduate school, they are less likely to stay in science and engineering. If they work, they are less likely to work in science and engineering.

In summary, Xie and Shauman

- Did not find that the "mathematics gap" is important.
- Found that career processes are fluid and dynamic.
- Found that being married and having children put women at a disadvantage.
- Found that sex differences in research productivity decline and can be attributed to differences in personal characteristics and structural features of employment.

Let me just emphasize this point: we have a temptation to try to find a single, simple explanation. There are two tendencies in finding simplistic explanations. Some scholars claim that everything is biology. Others claim that everything is discrimination. I think we should give up the naive idea that there is a single explanation.

—*Yu Xie*

SELECTIONS FROM THE QUESTION AND ANSWER SESSION

DR. KAMINSKI: Hello, I'm Deborah Kaminski from Rensselear Polytechnic Institute. I was fascinated by your idea that people could inadvertently decide not to perform as well on an exam because of the stereotypes. I'm wondering if that happens at the genius level as well? Perhaps there are so few women in the genius category, because genius is supposed to be male.

DR. SCHMADER: Yes, that's a very good question. And part of the theoretical assumption is that these effects might be most profound or strongest for people who care the most about excelling in the domain. So, for those women who care the most, we should see the strongest effects. And if the women who care the most are the women who do the best, then it could explain why you see a gap at the highest levels of achievement. There haven't been studies that have systematically addressed that question.

DR. SPALTER-ROTH: Hi, I'm Roberta Spalter-Roth from the American Sociological Association. First, we have a poster at this meeting suggesting that there is a real relationship between productivity and motherhood, and that family leave policies are granted to those women who are already more productive. So, leave is given as a reward for productivity, rather than as it was designed, as a needs-based policy. Hence, it reinforces the cycle of high and low productivity.

The second thing is that implicit assumption test we took, I found that really strange. I'm not sure what the study does.

DR. BANAJI: Okay, fair enough. So, the first thing to point out is that the test has nothing to do with the accuracy with which you put things into the right category. We worked very hard to make it very clear what belongs where. To classify a name like Mary as female and Peter as male is not the hard part.

The test measures the difference in the time that it takes to make the association of gender group with a particular attribute in the first round [male-science, female-home], where it is mentally compatible in the directional stereotype. The second round is the less compatible one [female-science, male-home], because that's not the stereotypic association.

With this group, not unlike others, the time interval difference between the first and second round was 700 to 1,000 milliseconds. That is a large statistical effect. That difference in time was substantial. It's big enough that we don't need a computer to measure it, a sundial will do. That's how big these biases are. And it just shows something very simple, that two things that have come to be paired repeatedly in our experience are going to be responded to as if they were one.

DR. CHUCK: My name is Emil Chuck. I'm at Duke University and I'm also involved in the National Postdoctoral Association. One of the things we found is that expectations regarding mentoring are really, really important, whether it's graduate students, postdocs or undergraduates, and play a significant role in whether people want to remain in the sciences. What do you think would be an effective means of reversing implicit biases in academe?

DR. FISKE: One of the things that we find in our broader work on stereotyping is that when people are in positions of power, they are very vulnerable to making prejudiced decisions about other people, because there is no feedback and very few consequences. I would argue that what you need to do is to build in accountability, so that training people effectively is part of how people are evaluated. You need to build people's sense of being interdependent with their subordinates, and not just having total, absolute power over them. And you need to reinforce people's values so that they are fair and unbiased. These three kinds of things about the relationship and the accountability and the values do help to overcome some of the implicit biases.

DR. SCHMADER: It is really easy to set up mentoring programs that women are "forced" to be a part of. We have to be cautious that these programs are framed in a voluntary supportive way, as opposed to saying, we know you're going to be having problems, and so here's a mentor that can help you. Mentors are really important, but in some sense the presence and welcoming open kind of environment mentors create is maybe more important than having it done in a very institutionalized way that sends this subtle, subversive message that it's expected that you will need mentorship.

DR. BANAJI: From what we have learned, the most important thing that we conclude is awareness. And not awareness in the old-fashioned sense that we need to go through diversity training once a year to know how to behave. If environments matter—and we have shown that implicit biases can be shaped by something as simple as who you see in front of you—then the mode for changing behavior needs to be changed.

Frank Dobbin, a sociologist at Harvard, has written a paper in which he analyzes 800 organizations from the 1970s on, and looks to see what happens to the diversity of the workforce in companies after diversity training was implemented. It turns out that the workforce becomes less diverse. There are many different reasons why this might happen. There are people who argue that this could be a backlash. Others, like me, think that it's a sense of, we checked the box off. I went to diversity training. And as a result, you don't bring to bear that particular lesson when making evaluations, because training and evaluations are mentally separate, physically separate, socially and psychologically removed.

Bias needs to be thought of in the same as we think about our physical selves. We know that exercising a lot the day after Thanksgiving dinner is not going to be sufficient. The change in body shape comes from this slow, hard work. And I think that the removal of bias needs to be thought of in this kind of incremental way, rather than the single one-shot thing.

Frank Dobbin found that mentoring programs work somewhat, and networking works somewhat, unlike diversity training. But what really works is an ombudsperson whose job it is to hold people accountable, and to ask the questions, and then old-fashioned affirmative action.

DR. VOGT: Christina Vogt, National Academy of Engineering. We know stereotype threat exists also between groups of males. Some groups will threaten white males. There is always a pecking order.

DR. SCHMADER: Stereotyped threat effects are situational. What that means is that any one of us in this room could experience stereotype threat if put in the right type of situation. You only have to think about what kind of group membership you might have that in a certain context would make you negatively stereotyped. So, as was mentioned, white men can show lower performance on a math test if they are told that the purpose of the test is to compare how whites do relative to Asians. Stereotype threat is a contextual effect, it is just that for women and minorities the context is more often chronically present.

DR. MANDULA: Barbara Mandula, EPA. This is a question for Toni about one of her early graphs. What you didn't mention was that men's performance went down when the task became an exercise rather than a test.

DR. SCHMADER: Members of advantaged groups can get some benefit from positive stereotypes. In any given individual study it often appears that men, when told the task is a lab exercise, suffer some performance decrement, or at least their performance is higher when they think it's an intellectual test. There is a meta-analysis that suggests that overall that effect is reliable, but the effect size isn't nearly as large as the threat effect that we see for women.

DR. GROSZ: Barbara Grosz, Harvard University, and also on the committee. I have a question of clarification for Yu Xie. You said that the greatest drop in participation was between high school and the bachelor's degree. But that conflicts with what other data I know that at the in the life sciences, women are more than 50% of the undergraduate students at many schools.

DR. XIE: The results that I presented were based on old data of all sciences. So, it is true that especially in recent years women's representation in biological sciences has been pretty high. In this particular analysis, it was a combined definition of science and engineering.

PANEL 3
ORGANIZATIONAL STRUCTURES

Panel Summary

Moving Beyond the "Chilly Climate" to a New Model for Spurring Organizational Change
Joan Williams, Center for WorkLife Law, University of California, Hastings College of the Law

Economics of Gendered Distribution of Resources in Academe
Donna Ginther, Department of Economics, University of Kansas

Bias Avoidance in the Academy: Challenges, Opportunities, and the Value of Policies
Robert Drago, Departments of Labor and Women's Studies, Pennsylvania State University

Gendered Organizations: Scientists and Engineers in Universities and Corporations
Joanne Martin, Graduate School of Business, Stanford University

Selections from the Question and Answer Session
Moderated by committee member ***Lotte Bailyn***

PANEL SUMMARY

The panel examined how the features of organizations, their rules, and their policies interact with gender to impose unequal demands or requirements on women.

Joan Williams discussed a new model for examining gender bias against women in academe that moves beyond the traditional concept of a "chilly climate." This model aims to describe in concrete terms the unrecognized patterns of stereotyping that negatively affect women in academe, to train people to recognize this bias for what it is, and to highlight an important new trend in federal employment lawsuits of which employers must be mindful.

Williams explained that we must use new metaphors and specific descriptions when naming bias, because how an issue is framed affects how it can be dealt with. Calling bias "subtle," "unconscious," or "implicit" makes it difficult to hold people responsible for the bias. Calling bias "unexamined," on the other hand, places the responsibility on the person who holds the stereotype.

Williams discussed the well-known concept of the glass ceiling and then introduced an important new trend in employment discrimination law: the concept of the "maternal wall." Also known as "family responsibilities discrimination," the maternal wall penalizes mothers, potential mothers, and fathers who seek an active role in family care. Mothers who face the maternal wall experience gender stereotyping in the way jobs are defined, in the standards to which they are held, and in assumptions that are made about them and their work—for example, a man who is absent is assumed to be presenting a paper, whereas a woman who is absent is assumed to be taking care of her children. They also face negative competence assumptions—assumptions that they are less competent or committed than other workers. In light of such bias, the maternal wall often pits women against women—for example, when women without children fear that making way for mothers may reinforce negative stereotypes about all women. Fathers, too, Williams explained, suffer from family responsibilities discrimination: As compared to mothers, fathers who take a parental leave or even a short leave to deal with family matters often receive fewer rewards, lower performance ratings, and are viewed as less committed.

Williams concluded by discussing the federal employment laws under which employees can sue—and employers can be sued—for maternal wall discrimination, including Title VII of the Civil Rights Act of 1964, the Pregnancy Discrimination Act, and the Family and Medical Leave Act. In sum, Williams argued, it is time to move beyond talking about what is actually gender bias as merely a "chilly climate" for women in academe. She argued for the need to create a new model for spurring institutional change that specifically names and identifies unexamined bias and considers the risk of family responsibilities discrimination lawsuits.

Donna Ginther examined the economic aspects of female academic careers, noting that a salary gap exists between male and female senior science professors and that marital status and parental status are major factors in determining career outcomes for women scientists. Ginther emphasized the importance of disaggregating the data by field and rank and of placing gender differences in a broader context. Women's representation in science varies by field: significant numbers of women are in the life sciences, much smaller numbers are in the physical sciences and engineering, and over half of the doctorates in the social sciences, except for economics, now go to women.

The percentage of tenured faculty who are female has lagged behind the percentage of women who earn doctorates in all fields. This gap may result from gender differences in hiring, in obtaining tenure, or both. In the social and life sciences, being female significantly and negatively influences women's chances of being in tenure-track jobs within 5 years of earning the PhD. Like Xie, Ginther found family status a highly significant factor in determining career progression: single women scientists were 16% more likely than single men to be in tenure track jobs 5 years after the PhD, and married women with children 45% less likely than married men with children. Marriage has a positive and significant impact of 22% on men getting a tenure track job whereas the effect of marriage for women is much smaller. Children, especially young children, significantly decrease the likelihood of women obtaining a tenure track job between 8% to 10% in all science fields, while having no significant impact on men. Ginther attributes these differences to the coincident timing of the tenure and biological clocks and women's role as the primary caregiver for children.

Ten years past the PhD, women faculty in engineering and the life sciences are marginally more likely than men to be promoted to tenure, but in other fields, female promotion is less likely. A significant salary gap exists between men and women at the full professor level but not at other ranks. Neither differences in family status nor productivity explain that discrepancy, nor does the imperfect competition in the academic labor market. The general pattern is consistent with the model that suggests that male advantage accumulates in the scientific world, with men consistently receiving greater rewards than women for accomplishments. Further research—with better data on discrimination and on scientists' research, resources, job prestige and other factors—is needed, Ginther said.

Echoing themes from Williams' talk, Robert Drago discussed the pervasive bias against caregiving that exists in many academic institutions and the strategies that academics use to try to prevent it from damaging their careers. "Productive bias avoidance" involves finding ways to minimize family commitments. The most obvious method is to have no children, and female academics indeed do have fewer children than members of other professions, such as female doctors or lawyers. Some 17% of women at research universities stay single, as opposed to 10% of men. In addition, 30% of women but only 13% of men have limited the number of their children to avoid career damage; 18% of women but 8% of men have delayed their second child for the same reason. Given the long periods of training in many sciences, that often pushes the second child into the mother's forties.

"Unproductive bias avoidance" involves efforts to deflect attention from one's family responsibilities and is "a new source of gender inequity." At research universities, more women than men decline to reduce their workload or to take needed parental leave to care for family, they miss children's events, and they return to work earlier than they desire after the birth of a child.

Joanne Martin examined how ostensibly gender-neutral organizational practices can disadvantage academic women. The 7- to 10-year tenure clock often imposes a severe conflict with the biological clock that is limiting women's reproductive years. Requirements to travel, to relocate, and to work long days are often more difficult for women, particularly those with family responsibilities. Performance evaluations based on subjective criteria often yield biased assessments. Exercising significant leadership is often more problematic for women because traditional feminine behavior is judged as "not tough enough," but assertive behavior inspires dislike.

The traditional approach that universities have used to open careers to women has been to merely hire women, Martin said. That has been presumed to give woman an "equal opportunity" to succeed. Because of gendered requirements and cultures of supposedly gender-neutral organizations, however, it produces high female attrition at every level, leaving only a handful of pioneers who manage to reach the top. Those pioneering women suffer problems including isolation, extreme visibility, unreliable feedback that is either too positive or too negative, and feelings of inauthenticity, which are especially severe for women in minority groups. The classic approach to these problems is to "fix the woman," but a more effective approach is to tailor responses to the characteristic issues produced at tipping points. Institutional interlocks among numerous organizations, such as families, schools and employers, require a coordinated effort and intra-organizational interventions to remove gender burdens.

MOVING BEYOND THE "CHILLY CLIMATE" TO A NEW MODEL FOR SPURRING ORGANIZATIONAL CHANGE[48]

Joan Williams
Center for Work-Life Law, University of California, Hastings College of the Law

Joan Williams discussed a new model for spurring organizational change that moves beyond the concept of a "chilly climate" for women in academe to identify unexamined bias and consider a new trend in federal employment discrimination lawsuits.

---

[48]For more detail, figures, and references, see the paper by Joan Williams in Section 2.

The challenge in science, as expressed by earlier speakers, is that gender bias "[does] not look like what we thought discrimination looked like."[49] The traditional language for talking about the position of women in science calls for eliminating the chilly climate by creating a culture of faculty support.[50]

In fact, the chilly climate often stems from documented patterns of gender stereotyping, some of which is outright illegal. The Center for WorkLife Law[51] (which Williams founded and directs) proposes a new model for creating institutional change. This model aims: first, to describe in readily understandable terms the patterns of stereotyping that create the chilly climate; second, to teach people to spot bias as it is happening; and third, to highlight the importance of a new trend in federal employment law of which institutions should be mindful.

In addition to glass ceiling discrimination and sexual harassment, is a trend called the "maternal wall" or "family responsibilities discrimination" (FRD), which penalizes mothers, potential mothers, and fathers who seek an active role in family care. The Center for WorkLife law has documented over 600 of these cases. One of the things emerging in these maternal wall cases is an alternative to the traditional way of proving discrimination. Traditionally, you would prove discrimination through use of a comparator, comparing the woman to a similarly situated man. But two recent maternal wall cases have had extraordinarily important holdings: one held that discrimination cases may also be proved through stereotyping evidence, even if you don't have a comparator;[52] another case said that cognitive bias—in that case attribution bias—was recognized as a form of stereotyping.[53]

Given the new importance of stereotyping evidence in discrimination law, we have thought a lot about issues of framing. We have heard some of the traditional language this morning—"subtle," "unconscious," "implicit" bias. Some of that language is not particularly helpful in the legal context. First of all, if this bias is so subtle, is it fair to hold people responsible legally? Secondly, if it's unconscious, how can it meet the standard for intentional discrimination? "Implicit" doesn't have those problems, but it's not sufficiently transparent really for use either in public education or certainly in the courtroom.

---

[49]Massachusetts Institute of Technology (1999). A study on the status of women faculty in science at MIT. *The MIT Faculty Newsletter* 11(4):14-26, http://web.mit.edu/fnl/women/women.html.

[40]Stanford University (1993). Report on the provost's committee on the recruitment and retention of women faculty. M. Strober, Chair.

[51]The Center for WorkLife Law is housed at UC Hastings College of the Law; http://www.worklifelaw.org.

[52]*Back* v. *Hastings-on-Hudson*, 365 F.3d. 107 (2d Cir. 2004).

[53]*Lust* v. *Sealy Inc.*, 383 F.3d 580 (7th Cir. 2004).

> *The new terminology that we have suggested is the terminology of "unexamined" stereotyping. Note how "unexamined" shifts the burden of proof. If it's unconscious, "oh my gosh, I didn't know." But if it's unexamined and you are clueless, whose fault is that? This new terminology also highlights that although stereotype activation is automatic, as Mahzarin Banaji pointed out this morning, stereotype application can be controlled.*
>
> —Joan Williams

WorkLife Law uses the law proactively to spur institutional change and organizational change by influencing intermediaries. In this case, human resource professionals are extremely important. This process is already underway with respect to the maternal wall. For example, one management side firm advised employers not only to avoid stereotyping, which is what the cases required, but also to consider offering telecommuting, flex time, and proportional pay and benefits for part-time work.[54] Once the potential for legal liability is established, often intermediaries institute the norms in a quite sweeping way.[55] But to use this new legal trend to spark organizational change, these stereotyping patterns must be easy to spot.

We all know about the *glass ceiling*. The glass ceiling penalizes women simply because they are women, and it does so in two distinct ways. Some of the patterns make it harder for women to be perceived as competent, which, of course, makes it harder to succeed. For example, when women are judged on accomplishments, but men are judged on potential; performance evaluations are gender-biased; double standards are applied to men and women; women must be superstars to survive while men can be average; women are kept out of the loop; jobs are defined in terms of masculine patters; or women must play certain roles in order to accepted. The other patterns penalize women for being too competent, which again makes it harder to succeed. For example, when women are considered aggressive, while men are considered assertive, but women are also penalized for not being aggressive enough; women are considered shameless self-promoters, while men are considered to know their worth; successful women are sexually harassed.

In addition to the glass ceiling is the *maternal wall*, which penalizes mothers, women perceived to be potential mothers (which is often most women), and also

---

[54]TP Krukowski, SC Costello (2002). Discrimination: A glass ceiling for parents? *Washington, DC Employment Law Letter* 3(6):1, http://www.hrhero.com/dcemp.shtml.

[55]EE Kelly and F Dobbin (1999). Civil rights law at work: Sex discrimination and the rise of maternity law policies. *American Journal of Sociology* 105:455-492; LB Edelman (1997). Legal ambiguity and symbolic structures: Organizational mediation of civil rights law. *American Journal of Sociology* 97(6):1531-1576; R Stryker (2003). Mind the gap: Law, institutional analysis, and socio-economics. *Socio-Economic Review* 1:335-367.

fathers who seek an active role in family care. That's why the more technical name for the maternal wall is family responsibilities discrimination. It is linked to being the primary caregiver or providing care to family members.

> *Here is an extremely important demographic fact: 95% of mothers aged 25-44 work less than 50 hours a week year round. So, basically all you have to do is define full-time as 50 or more hours a week to come close to wiping mothers, and therefore three-quarters of women, out of your labor pool.*
>
> *—Joan Williams*

Maternal wall patterns of discrimination include: jobs defined around masculine patterns (for example, selecting workers who are "single-minded"); role incongruity (for example, she cannot be both a mother and a full-time academic); prescriptive stereotyping, whether benevolent or hostile (for example, she shouldn't worry about her work, but should just focus on her family); attribution bias (for example, an absent man is assumed to be presenting a paper, but an absent woman is assumed to be taking care of her kids); and leniency bias (for example, women are held to higher standards than men).

Another key component of maternal wall patterns of discrimination is negative competence assumptions about mothers. A 2005 study found that "relative to other kinds of applicants, mothers were rated as less competent, less committed, less suitable for hire, promotion and management training, and deserving of lower salaries."[56]

Another dynamic that is not very well understood is what Williams terms *gender wars*—tensions among women themselves. This is an extremely acute problem in academics, because 50% of women academics in science have no children. Many of these women are "child-free"—meaning that they do not want children. These women may feel anxiety about making way for mothers out of fear that having to accommodate mothers reinforces negative stereotypes about all women. On the other hand, many of these women are "child-less"—meaning that they want or wanted, but do not have, children. These women may think, "Why should she have it all, when I had to sacrifice so much?" Thus, the maternal wall very often pits women against women. It is important to recognize that this phenomenon is actually a *result* of gender discrimination, *not* proof that discrimination against mothers "is not a gender problem."

There is also family responsibilities discrimination against fathers. In one study, when compared to mothers, fathers who took parental leave were recommended for fewer rewards and viewed as less committed, and fathers with even a

---

[56]SJ Correll and S Benard (2005). Getting a job: Is there a motherhood penalty? Presentation at American Sociological Association Annual Meeting, August 15, 2005, Philadelphia, PA. http://sociology.princeton.edu/programs/workshops/Correll_Benard_manuscript.pdf.

short work absence due to a family conflict were recommended for fewer rewards and had lower performance ratings.[57] In academics this translates into what Robert Drago calls "unproductive bias avoidance"—for example, in the case of an untenured professor who told his mentor that he did not dare even to ask about parental leave, much less take it, for fear his career would be over.

People need to understand their rights as employees, and institutions need to understand the consequences of committing family responsibilities discrimination: potential lawsuits. Maternal wall cases have been brought under a number of legal theories in federal employment law, including the following:

- *Disparate treatment under Title VII of the Civil Rights Act of 1964*—for example, a female professor who was treated worse and subject to greater scrutiny by colleagues after she had a baby
- *Retaliation under Title VII of the Civil Rights Act of 1964*—for example, when a woman faces negative career consequences for protesting a denial of maternity leave or asking to stop the tenure clock while she is on maternity leave
- *Interference with rights under the Family Medical Leave Act (FMLA)*— for example, a female professor who was pressured to reduce the amount of time she took on maternity leave (In certain circumstances, the FMLA provides 12 weeks of unpaid leave and guaranteed reinstatement; one study showed that 40% of academic women surveyed returned to work from leaves earlier than they wanted to.[58])
- *Violation of the Pregnancy Discrimination Act (PDA)*—which protects employees from discrimination based on pregnancy, childbirth, and related medical conditions and requires that pregnancy be treated the same as other temporary disabilities.

According to a recent study,[59] over one-third of academic institutions had family or child rearing policies that probably violate the Pregnancy Discrimination Act (PDA). This, of course, places mothers in extremely awkward positions. They have to impose on their colleagues for leave that they should be entitled to, and they have to fight political battles to get that leave.

—Joan Williams

---

[57]CE Dickson (2003). The impact of family supportive policies and practices on perceived family discrimination, (dissertation).

[58]MA Mason (2003). UC Berkeley faculty work and family survey: Preliminary findings, http://universitywomen.stanford.edu/reports/UCBfacultyworknfamilysurvey.pdf.

[59]S Thornton (2003). Maternity and childrearing leave policies for faculty: The legal and practical challenges of complying with Title VII. *University of Southern California Review of Law and Women's Studies* 12(2):161-190.

In conclusion, Williams called for moving beyond talking about gender bias as merely a "chilly climate" for women in academe. She argued for the need to create a new model for spurring institutional change that specifically names and identifies unexamined bias and considers the risk of family responsibilities discrimination lawsuits against employers.

ECONOMICS OF GENDERED DISTRIBUTION OF RESOURCES IN ACADEME

Donna Ginther
Department of Economics, University of Kansas

Donna Ginther focused her comments on the economics of gender differences in employment outcomes in academia. She observes gender and race differences in employment outcomes. From the economics perspective, gender differences in employment outcomes result from a variety of factors besides discrimination.

- *Differences in productivity.* Are men more productive than women?
- *Differences in choices.* Women's choice of occupations and jobs affect their employment outcomes.
- *Imperfectly competitive markets.* Becker's theory of discrimination was predicated on perfect competition; however, universities are not perfectly competitive. In fact, they have monopsony power where universities act as single purchasers of academic labor and have more market power than employees.
- *Job matching.* This theory suggests that differential employment outcomes result from one group performing better on the job than another.

If none of those theories explains the employment-outcome difference, then what is left over could be attributed to discrimination. That said, economists, on average, do not believe that discrimination explains observed gender differences in employment outcomes.

> *There is no single scientific labor market. As a result, we need to disaggregate the data. We need to look at the different scientific labor markets because they have different outcomes for women. We need to make comparisons across fields to understand the status of women relative to one another. Hiring, salary, and promotion outcomes are interrelated. You cannot look at one without considering the others.*
> —*Donna Ginther*

> **BOX 1-3**
> **The Economist's Perspective**[a]
>
> ~Economists view the world as being organized by markets, and assume that markets matter. Thus, supply and demand determine employment outcomes.
> ~Economists assume that equally productive workers will be paid the same. Thus, we should not observe gender differences provided that men and women are equally productive.
> ~Discrimination exists, but market competition will remove it. In other words, if you have a perfectly competitive market, some employer can exploit the fact that it is not paying women enough, hire only women, and then become more profitable.
>
> ---
> [a]Gary Becker won the Nobel prize in economics in part for his theories of discrimination. GS Becker (1971). The Economics of Discrimination. Chicago: University of Chicago Press.

What explains the differential employment outcomes in science and engineering fields? To examine hiring, promotion, and salary, Ginther used the 1973-2001 waves of the Survey of Doctorate Recipients (SDR).[60] Because the SDR is longitudinal, respondents can be tracked over time. She split the data into fields: life sciences (agriculture, food science, and biology), physical sciences (chemistry, earth sciences, physics, and mathematics), engineering, and social science (economics, psychology, sociology, anthropology, and political science). Control variables include the demographic variables of gender, race, and age; academic field and degree; rank and tenure status; and institutional characteristics (Carnegie rankings, public or private). She included control variables for primary work activities which indicate whether the respondent primarily teaches, does research, manages, or engages in another activity. She also included an indicator for whether a respondent receives government support and some measures of publications.

Ginther's research shows again that women's representation depends on field (Figure 1-7). Since the 1970s there has been tremendous growth in the number of doctorates awarded to women. In the physical sciences, there is still anemic representation of women, but in the life sciences and the social sciences (except in economics) half or more of doctorates are awarded to women.

---
[60]From 1987 to 1995 the SDR also followed people in the humanities; for these years, Ginther includes humanities in her analysis.

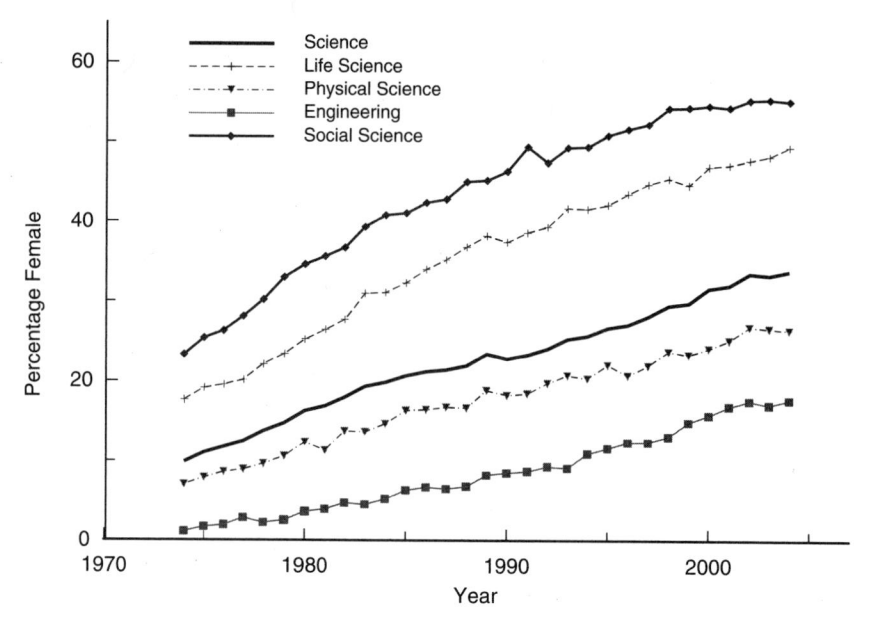

FIGURE 1-7 Percentage of doctorates granted to females.
SOURCE: National Science Foundation (1974-2004) *Survey of Earned Doctorates.* Arlington, VA: National Science Foundation.

However, even in life science and social sciences, the percentages of women tenured is low. For example, in social science, over 50% of the doctorates have been women since 1990, but in 2001 only 28% of tenured faculty were women (Figure 1-8).

Does that discrepancy result from differences in hiring or from differences in promotion? Ginther examined gender differences in tenure-track jobs within 5 years of earning a PhD and measured the effect of being female on getting a tenure-track job. She found that single women are significantly more likely than single men—by 11 to 21%—to have tenure-track jobs. Marital status and presence of children drive this result and explain the leaky pipeline.

Marriage has a positive and significant impact of 22% on men getting a tenure-track job whereas the effect of marriage on women ranges between 0 and 8% for all science, life science, and social science fields. Children, especially young children, significantly decrease the likelihood of women obtaining a tenure-track job between 8 to 10% in all science fields, life science, and social science while having no significant impact on men.[61]

---

[61] MA Mason and M Goulden (2002). Do babies matter? The effect of family formation on the lifelong careers of academic men and women. *Academe* 88(6):21-27, http://www.aaup.org/publications/Academe/2002/02nd/02ndmas.htm.

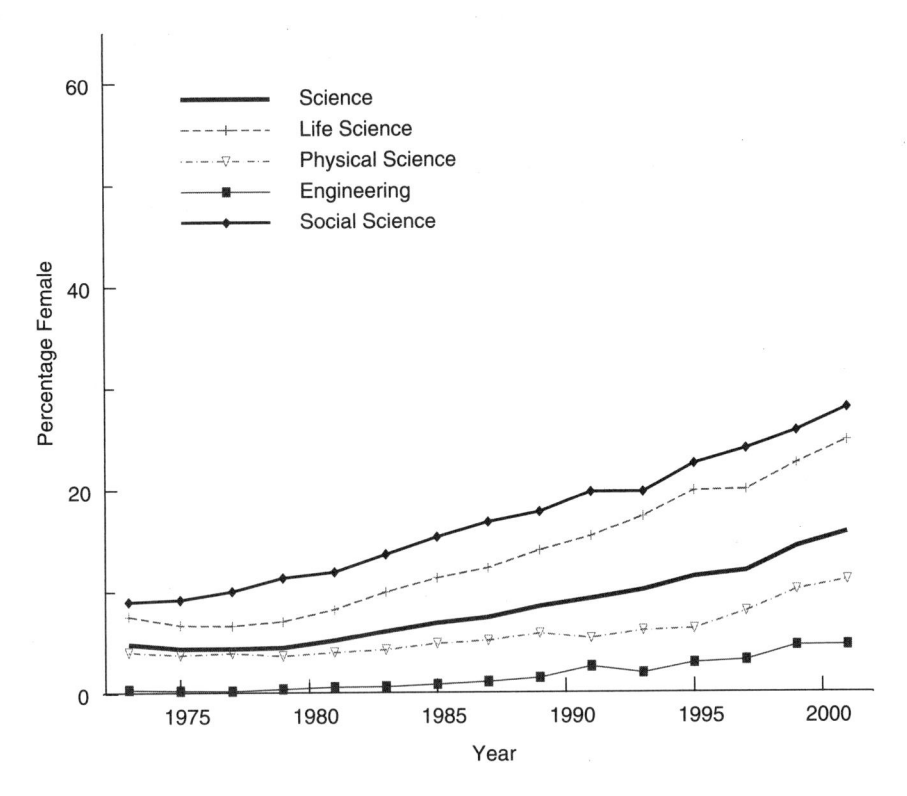

SECTION 1: SUMMARIES OF CONVOCATION SESSIONS   59

FIGURE 1-8 Percentage of tenured faculty who are women.
SOURCE: National Science Foundation (1973-2001). *Survey of Doctoral Recipients*. Arlington, VA: National Science Foundation.

The differential impact of marriage and children may be explained by a number of factors. Women may choose to have children instead of pursuing an academic career because of the coincident timing of the tenure and biological clocks. The dual career problem may also play a role. Career hierarchies in marriage often result in the husband's career taking precedence over the wife's career. If it is difficult to obtain two tenure-track jobs, she may choose to have children instead of investing in her career.

In particular, with respect to hiring policies, the dual career problem should be taken seriously. There is an economic advantage for a university to hire couples, because couples are less mobile. The university can probably keep them longer.
—*Donna Ginther*

Ginther also examined gender differences in promotion 10 years past the PhD. Overall, she found a 1.4% gender difference in promotion to tenure 10 years past PhD. The gender promotion gap varies significantly by field. In the social sciences, excluding economics, women are 8% less likely to have obtained tenure; in the life sciences, 2% more likely; in the physical sciences 3% less likely; in engineering, 4% more likely; in humanities, 8% less likely. Economics is the outlier, in which there is a 21% promotion gap in favor of men.

After examining hiring and promotion, Ginther considered the gender salary gap. In the economy as a whole, women earn 75 cents for every dollar a man earns. In engineering, women earn 80 cents for every dollar. Previous research has shown that if academic rank is factored in, the gender differences in salary go away, except for full professors. What is an 18% difference favoring men in science as a whole falls to just over 5% in science for assistant professors, even less for associate professors. For full professors there is a 13.2% salary gap. One-third of the 13.2% salary gap is attributable to valuing the observable qualifications of women differently than men. Across the campus in the humanities, there is essentially no salary difference at any level. Something is going on in the humanities and the social sciences relative to science. For some reason, there are huge salary discrepancies at the full-professor rank in the sciences but not in the social sciences or humanities.

What are the economic explanations for the salary gap? It is not the result of marriage and children, except in the life sciences. Women are more productive on the average than men at Research I institutions; productivity is not explaining the gap. The salary gap is probably not the result of monopsony in the academic labor market. We also can dismiss the explanation that women are not good scientists, because they would not be full professors if that were the case.

What I find is that the salary gap is explained largely by gender differences in work experience and that men are rewarded more than women. That is consistent with the cumulative advantage model.
—*Donna Ginther*

To address outstanding questions, Ginther recommended improving the quality of data. She suggested building on existing datasets, including the SDR and the National Institutes of Health Consolidated Grant File. The postdoctoral process should be examined because it seems to be a key point at which women are dropping out. In particular, the SDR should add questions on publications and citations, grant awards, laboratory space, number of graduate students supervised, and a special module on postdoctorates. She called for additional questions on spouses—their education, their employment, their earnings, and how much child care time is allocated. She also urged universities to undertake a systematic review of academic salaries.

BIAS AVOIDANCE IN THE ACADEMY: CHALLENGES, OPPORTUNITIES, AND THE VALUE OF POLICIES

Robert Drago
Labor Studies and Women's Studies, Pennsylvania State University

Robert Drago discussed caregiving bias avoidance[62] in the academy. That motherhood is a serious problem for professional advancement in academe is shown in Figure 1-9. The data show the percentages of women faculty, doctors, or lawyers with a baby (0-1 yrs old) in the household. The women academics are having fewer babies than the doctors or lawyers. Academe is obviously a tough sector to work in.

Another study[63] suggests that there is a bias against caregiving in the professional workplace. It affects women more than men, partly because of stereotyping, but anybody who exhibits symptoms of caring for family will be penalized or experience bias in the workplace, and that leads to the new glass ceiling, the *maternal wall*.

Drago and colleagues performed a series of focus groups with faculty parents at Pennsylvania State University and found evidence of caregiving bias. But there was more evidence of what they came to call *bias avoidance*. That is, faculty are smart enough to figure out that they are going to run up against biases, so they find strategies to avoid them.

They categorized two types of strategies: productive and unproductive. In *productive bias avoidance* you find ways to minimize family commitments to create more time for career—having fewer children than you wanted, delaying having children, buying wife-replacement services, and so on. These strategies are productive, because they free up time. Productive bias avoidance may be efficient, but it is inequitable. That is, it is not distributed equally across genders, so it is not fair. In *unproductive bias avoidance*, you ignore your family commitments, which is both inefficient and inequitable. Unproductive bias avoidance has no general rationale and is a game that has unknown rules. That is, you can not ask, Will I experience bias against caregiving if I have a child? To do so, you have to admit that you care about children, and this might potentially write off your career.

With those initial results, Drago and colleague Carol Colbeck did a national study of faculty in chemistry and English at 507 schools, including all the

---

[62]Kathleen Christianson at the Sloan Foundation coined the term bias avoidance, and it comes out of research by Joan Williams.
[63]JC Williams (2001). *Unbending Gender. Why Family and Work Conflict and What to Do About It*. New York: Oxford University Press.

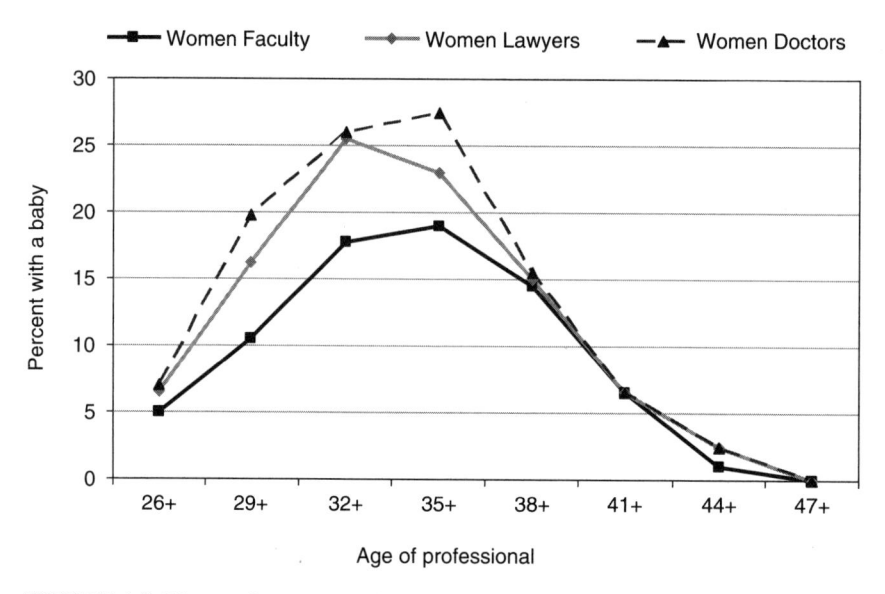

62   COMPONENTS OF SUCCESS FOR WOMEN IN ACADEMIC SCIENCE & ENGINEERING

FIGURE 1-9 Women fast-track professionals with babies in the household, by age of professional.
SOURCE: US Census, 2000 Public Use Microdata 5% Sample, prepared by M Goulden.

BOX 1-4
Bias Avoidance Behaviors[a]

Productive Bias Avoidance	Men	Women
Stayed single to achieve academic success.	10 %	17 %
Limited the number of children—that is had fewer children than desired—to achieve academic success.	13 %	30 %
Delayed having a second child until after tenure.[b]	8 %	18 %

Unproductive Bias Avoidance	Men	Women
Did not take a reduced load when needed for family commitments.	19 %	30%
Did not take parental leave even though it was needed.	27 %	31 %
Missed some of the young children's important events, because wanted to be taken seriously.	34 %	40 %
Came back too soon after a new child.	12 %	46 %

---

[a]Survey results presented are restricted to Research I institutions.
[b]Given that on average, if US women are 34 when they receive their PhD, that puts the second childbirth in a woman's forties. Less than 1% of all live births are to women over the age of 40 even today. So this strategy does not always work.

Research I universities.[64] They did 10 case studies and shadowed 13 faculty for about 650 hours. They asked subjects whether they engaged in avoidance behaviors and if so, whether for career success or to appear committed or to be taken seriously (Box 1-4).

Avoidance behaviors are distributed unevenly. They appear to be a new source of gender inequity not seen in salary or promotion figures. Women are having to engage in these largely hidden behaviors. They can not ask anyone whether they need to engage in the behaviors, because of how the "game" is structured. Bias avoidance more often affects women and is reduced by supervisor support.[65] After introducing a control for positive affect, Drago found that upbeat, happy people engage in bias avoidance behaviors less often.

What are the returns to avoidance behaviors? Are people getting tenured earlier, or were they reducing the time between PhD and the point of tenure? For women who engaged in bias avoidance behaviors, time to tenure was reduced by productive bias avoidance behavior; the age at tenure was also significantly reduced.

There are payoffs for avoidance behaviors. With productive bias avoidance, that is no surprise. You are making more time by not having children or delaying having them. You should get tenure earlier and move through more quickly. The real surprise is that unproductive bias avoidance behaviors—which do not free up any time and may even be increasing the burden of trying to handle work and family—for men reduced the time to tenure by over a year and for women reduced the age at tenure by over a year. Playing the game has a payoff.
—Robert Drago

To address the question of whether institutional policies reduce the incidence of bias avoidance, Drago matched the Mapping Project data to a survey of 250 schools' work-life policies performed by Carol Hollenshead and Beth Sullivan. The policies included paid maternity leave, reduced hours, child and elder care, flexible hours, and some connection between policies for faculty, staff, and students. Drago created a scale out of the eight bias avoidance behaviors and correlated

---

[64] R Drago, C Colbeck, KD Stauffer, A Pirretti, K Burkum, J Fazioli, G Lazarro, and T Habasevich (2005). Bias against caregiving. *Academe*. Sept/Oct. http://www.aaup.org/publications/Academe/2005/05so/05sodrag.htm. *Research I university* was a category formerly used by the Carnegie Classification of Institutions of Higher Education to indicate those universities in the United States which received the highest amounts of Federal science research funding. The category is, since 2000, obsolete, but the term is often still used.

[65] R Drago, C Colbeck, KD Stauffer, A Pirretti, K Burkum, J Fazioli, G Lazarro, and T Habasevich (2005). Bias against caregiving. *Academe*, http://www.aaup.org/publications/Academe/2005/05so/05sodrag.htm.

them with work-life policies. Negative correlations would indicate that with more work-life policies there is less bias avoidance. However, instead of negative correlations they found positive correlations between policies and bias avoidance for not taking parental leave, indicating that the more policies that existed, the less they were used.

I got to thinking, well, which universities have gotten on the Working Mother Top-100 list? Harvard, Stanford, and MIT. Those are the only three schools. They have great policies, but they are extremely tough places to work for parents. If we reexamined the data and focused on the subsample of women faculty at research universities, all fairly tough schools to work at, all of a sudden we start seeing the negative correlations that we expected.

—*Robert Drago*

GENDERED ORGANIZATIONS: SCIENTISTS AND ENGINEERS IN UNIVERSITIES AND CORPORATIONS

Joanne Martin
Graduate School of Business, Stanford University

Joanne Martin shifted the focus from documentation of discrimination to institutional change efforts.

The old-fashioned approach to changing gender inequality was to hire more women, who would supposedly have equal opportunities to succeed. The problem was that this strategy was based on a false assumption: that organizational structures and cultures are gender-neutral. We know they are not. Many things that look gender-neutral, like the requirement that you have to travel and present your research or geographically relocate,[66] are tougher for women on the average than for men, particularly for those women with caregiving responsibilities.

—*Joanne Martin*

---

[66]AR Hochschild (1997). When work becomes home and home becomes work. *California Management Review* 39(4):79-97. Other practices include tenure-biological clock conflicts, subjective criteria in performance appraisals (C Wenneras, and A Wold (1997). Nepotism and sexism in peer-review. *Nature* 387:341-343; M Heilman, AS Wallen, D Fuchs, and MM Tamkins (2004). Penalties for success: Reactions to women who succeed at male gender-specific tasks. *Journal of Applied Psychology* 89:416-427, long hours, and acceptance of women in leadership roles (KH Jamieson (1995) *Beyond the Double Bind*. New York: Oxford University Press).

> **BOX 1-5**
> **Pioneers Have Predictable Problems**[a]
>
> —Exclusion and isolation.
> —Extreme visibility, particularly for woman of color. Every failure and every mistake is public. The good news is, if you succeed, that also is highly visible.
> —Unreliable performance feedback. Like flying an airplane without a gyroscope, you cannot rely on the feedback you are getting, and so you rely on your own sense of where you are doing well and where you are doing poorly.
> —Very strong probability of being unfairly promoted and unfairly paid.
>
> ---
>
> [a]RM Kanter (1977). *Men and Women of the Corporation.* New York: Basic Books; J Crocker and KM McGraw (1984). What's good for the goose is not good for the gander: Solo status as an obstacle to occupational achievement for males and females. *American Behavioral Scientist* 27(3):357-369.

The first-stage remedy of *add women and stir* is often unsuccessful in part because organizational structures and policies are gendered, and women leave more than men at each level of the promotion hierarchy.[67] At the highest levels, only women pioneers remain. This will be the situation for years and years for women scientists and engineers.

In addition to the problems faced by pioneers, another problem, which has received less attention in the literature, is women's discomfort in male-dominated cultures.[68] Inauthenticity problems are heightened when one is a woman of color.[69] The result is that women, particularly women of color, quit.

The classic remedy is to fix the women. Training women to be better leaders, to be more assertive, and to have the same kinds of tough

---

[67]J Acker (1990). Hierarchies, jobs, bodies: A theory of gendered organizations. *Gender and Society* 4(2):139-58; AJ Mills and P Tancred (1992). *Gendering Organizational Analysis.* London: Sage Publications.

[68]J Martin and D Meyerson (1998). Women and Power: Conformity, Resistance, and Disorganized Coaction. In *Power and Influence in Organizations*, Eds. RM Kramer and MA Neale. San Francisco: Sage Publications.

[69]AM Morrison (1992). New solutions to the same old glass ceiling. *Women in Management Review* 7(4):15-19; E Bell and SM Nkomo (2001). *Our Separate Ways: Black and White Women and the Struggle for Professional Identity.* Boston: Harvard Business School Press.

> *negotiating skills that men have encourages women to act like men and leaves intact the inauthenticity problem. Such training programs may be needed in the short term, but if inauthencity feelings are a real reason why women are dropping out, fixing the women won't fix the problem.*
>
> —*Joanne Martin*

A better approach is to study the kinds of problems that occur as the percentage of women in an occupation changes. Those kinds of problems are actually pretty well understood by organizational researchers.[70] There are three tipping points where the nature of the problems experienced by women changes drastically. The first, and the one that is most relevant in our lifetimes for women who are scientists and engineers, is the point at which the glass ceiling starts to break, and women start to enter into higher level positions, for example, department chair, dean, or college president. At 18-22% female, there is a critical mass. Women start to get leadership positions, and sometimes they get together, discuss their common interests, and organize collectively, for example for better work-family policies. So far, so good; but there is also a problem with this first tipping point: it is the first point at which white male backlash starts to appear. Women are starting to be a threat; as a result, they get resistance, sometimes overt resistance, from men.

The second tipping point is a temporary nirvana. When you have 40-60% women in an occupation, gender issues tend to disappear. People simply don't worry about gender issues, and evidence of discrimination and unfair pay and promotion policies are gone. It is a temporary nirvana; it doesn't last long. It quickly—amazingly quickly—switches to the third tipping point: occupational sex segregation.[71] That occurs when an occupation is either 90% female or 90% male.

The vast majority of occupations in the United States are sex segregated. In medicine, neurosurgery is still very close to being an all-male occupation, whereas pediatrics is beginning to be dominated by women. When Stanford did its big study in 1993, it found that almost all the hard sciences and engineering specialties were virtually all male, with one exception—biology. Occupational sex segrega-

---

[70]TF Pettigrew and J Martin (1987). Shaping the organizational context for Black American inclusion. *Journal of Social Issues* 43(1):41-78; M Gladwell (2000). *The Tipping Point: How Little Things Can Make a Big Difference.* Boston: Little, Brown.

[71]MH Strober and C Arnold (1987). The Dynamics of Occupational Segregation Among Bank Tellers. In Eds. C Brown and J Pechman, *Gender in the Workplace.* Washington, DC: Brookings Institution; F Conley (1998). *Walking Out on the Boys.* San Francisco: Farrar, Straus and Giroux; Stanford University Committee on Recruitment and Retention of Women, 1993; JC Touhey (1974). Effects of additional women professionals on ratings of occupational prestige and desirability. *Journal of Personality and Social Psychology* 29:86-89.

tion is a crucial issue because, in the United States at least, female jobs have lower pay and lower prestige.

We need to tailor and plan organizational change programs to fit the tipping-point stage. What you do for a biology department is going to be very different from what you do in a medical school which in turn will depend on whether you are working on the neurosurgeons.
—*Joanne Martin*

There are standard things that you do in an organization when you are just starting to try to get the ball rolling: You start counting things—not just personnel and pay, but also the number of square feet of a laboratory space.[72] Networking and mentoring are also classic strategies at this stage. Access to effective mentors is very important, particularly for minority-group women.[73]

Next, we need to de-gender organizations. The change strategies here are a little more ambitious. First is the 7-10 year tenure clock. Scientists and engineers in academe and corporations are under similar time demands. We know that the tenure clock and the fast-track schemes are harmful to women because they conflict with the biological clock, that is, the number of years that women have to have children before the danger of birth defects gets large.

Young women are telling us that they do not want to be like the pioneer generation, the super women who have gotten very tired trying to balance work and family. For example, at the Harvard and Stanford business schools, 70% of the female MBAs who graduate stop out of the workforce; of the female business school graduates from the Harvard classes of 1981, 1986 and 1991, only 38% are still working full time today.[74] Most who stop out would like to re-enter the workforce, but they are handicapped by having that hole in their CVs. They say that they want us to change organizations and change university curriculum so they can brush up their skills and re-enter the workforce after they stop out. That suggests that we think about instituting, for example, postdoctoral fellowships to facilitate re-entry. We need to revamp career tracks in academe, as well as in business.[75] That is going to be a long, hard struggle.

---

[72]Massachusetts Institute of Technology (1999). A study on the status of women faculty in science at MIT. *The MIT Faculty Newsletter* 11(4):14-26, http://web.mit.edu/fnl/women/women.html.

[73]SD Blake-Beard (2001). Taking a hard look at formal mentoring programs: a consideration of potential challenges facing women. *Journal of Management Development* 20(4):331-345; AJ Murrell, F Crosby, and R Ely (1999). *Mentoring Dilemmas: Developmental Relationships Within Multicultural Organizations*. Mahwah, NJ: Erlbaum; KE Kram (1985) *Mentoring at Work: Developmental Relationships in Organizational Life*. Glenview, IL: Scott Foresman.

[74]See M Conlin, J Merritt, and L Himelstein (2002). Mommy is really home from work. *Business Week* (November 25), http://www.businessweek.com/magazine/content/02_47/b3809108.htm

[75]SA Hewlett and CB Luce (2005) Off-ramps and on-ramps: Keeping talented women on the road to success. *Harvard Business Review* 83(3):43-46, 48, 50-54 passim.

Next, we have individual *tempered radicals*,[76] people who have worked from within an organization for change that benefits women. What we need to do is figure out what tempered radicals have done that has worked, share those strategies, and use them to train the next generation of tempered radicals. There are wonderful strategies out there, but there are few opportunities for us to share that knowledge and pass it on.

Finally, and most exciting, is the research on *small-win experiments*. Small wins are not small at all. The idea is that you go into an organization and implement a very small-scale experiment that, if it worked, would shatter gender stereotypes and open doors to equality as they had never been open before. Meyerson and Fletcher went to the Body Shop, the cosmetics company in England. The lipstick factory employed people on the assembly line in feminine uniforms, and every assembly line had a male supervisor in a white lab coat. The experiment abolished the male supervisor with the lab coat on two of the assembly lines, and had the assembly line workers, who were all women, rotate in and out of the leadership position. Productivity skyrocketed, and at the end of the experiment every woman on that assembly line had had leadership experience, and two or three of them actually got promoted. That's a small win that matters. Lotte Bailyn did the same thing at Xerox and changed the norms about working impossibly long hours.[77]

Because of interlocking institutions, which were mentioned previously by Yu Xie, in academe more than in any other domain, we can not just change universities. We have to change all kinds of other institutions simultaneously. And in designing change programs, you can not duck the need to look at families and the need to look at employing organizations more broadly.

As to future research, we should go beyond counting bodies and providing statistics, beyond documenting unfairness. That does not change behavior. What is needed are organization-level change programs tailored to the tipping points, because outcomes will not change until the process changes. For example, the first tipping point is probably the most germane for science and engineering. Once an organization or department is 18% women, what helps women to see their common interests, to network, and to organize collectively? What kinds of programs work best for minority women? How can male backlash be minimized? And when should policies benefit both sexes, and when should they not?[78]

---

[76]DE Myerson and MA Scully (1995).Tempered radicalism and the politics of radicalism and change. *Organization Science* 6(5):585-600.

[77]R Rapoport, L Bailyn, JK Fletcher, BH Pruitt (2002). *Beyond Work-Family Balance: Advancing Gender Equity and Workplace Performance.* San Fransisco: Jossey-Bass; D Meyerson (2003). *Tempered Radicals: How People Use Difference to Inspire Change at Work.* Boston: Harvard Business School Press.

[78]H Ibarra (1992). Homophily and differential returns: Sex differences in network structure and access in an advertising firm. *Administrative Science Quarterly* 37(3):422-447; DA Thomas and JJ

At the institutional level of analysis, Williams called for large scale, cross-institutional change programs that go beyond the boundaries of a single organization or even a single institution; where a single institution would be a member of class of organizations like "high schools" or "universities" or "families." How can the institutional interlock of families, schools, and religious organizations be broken so that gender progress is not blocked?[79] Finally, Williams said a lot could be learned by studying other countries. An example was the Australian equal opportunity agency (EOWA), which has given financial awards to organizations, including universities, for exemplary practices. It also has software on line that can be used to calculate whether men and women are being paid fairly.

SELECTIONS FROM THE QUESTION AND ANSWER SESSION

DR. CHUCK: Emil Chuck, Duke University. Certainly, many of these ideas are great, but how much money is it going to take? That has been the question that has confronted our parents group over at Duke University and our advocates for the last three years when it comes to human resources, finding less expensive child care, and so forth.

DR. MARTIN: You need to count the lost cost in faculty time, and in all other kinds of time, when you hire women who then quit.

DR. WILLIAMS: And in the sciences, the business case can be really robust when a university pays $200,000 for example, to hire a single person who will then leave because of work-family issues. With respect to graduate students, what graduate students very often need is part-time inexpensive childcare. Childcare, like everything else, tends to be conceptualized in an on/off model. If you had part-time child care slots through a co-op system, that might actually be far better for graduate students than an extremely expensive subsidy per child care slot, which is the classic model.

DR. DRAGO: Many universities, including Penn State, have networks of undergraduate child care and babysitting at some level. So, you can connect, and you can usually find somebody. The second is there is federal money through CCAMPIS grants,[80] which many universities have used to fund student child care. Third, you have to be organized. We have been fighting to get a third childcare center at Penn State for a while. It is just going to take time, and you have to keep

---

Gabarro (1999). *Breaking Through: The Making of Minority Executives in Corporate America.* Boston: Harvard Business School Press.

[79]J Martin (2006). *Gender Equity Interventions: What Works, What Doesn't, and Why.* Manuscript in preparation. Stanford University, Stanford, CA; L Wacquant (1997). For an analytic of racial domination. *Political Power and Social Theory* 11:221-234.

[80]The Child Care Access Means Parents in School (CCAMPIS) grants are administered through the Department of Education. See http://www.ed.gov/programs/campisp/index.html.

publicizing it. Fourth, I would say without federal money we're never going to solve the childcare problem. It's too big.

DR. GINTHER: A study of private sector employers found that only 2% provide paid parental leave. Coherent parental leave policies are really hard to come by even in academia. Joan Williams mentioned Saranna Thornton's work where one-third of universities had illegal family leave policies.[81] I think the first thing is to get universities to be law abiding institutions.

DR. SIMPSON: Carol Simpson, Worcester Polytechnic Institute. I'm at an institution where we are fortunate in having reached the first tipping point, and it's predominately an engineering institution, so we are very pleased with that. But, what that means is now we have two female faculty members in one department, three in another, one in another. And there is a sense of real isolation among those women. One strategy that I have used as provost that seems to be bearing some fruit is to bring faculty together at my invitation—in informal receptions, two or three times a year. Still, while this has had a very positive impact, it has not solved the problem. We lost two women this year by attrition. There is still a lot of work to be done.

DR. CARNEY: Arlene Carney, University of Minnesota. I was wondering if anyone on the panel could address the existence of data about stopping the tenure clock? I find it very difficult to convince our young untenured faculty members to take advantage of our policies. Are there data that show that women who have stopped the tenure clock have suffered, or that tenure committees actively have misjudged people because of stopping the tenure clock?

DR. BAILYN: As many of you know, Princeton and MIT have made it stopping the clock automatic, so that people do not have to ask for it. We did have some data in the period before it was automatic, and that's where the evidence comes that people who took leave didn't get tenure. I don't think we have enough time yet to know whether the automatic extension is going to help. At MIT it's only been in place for five years. There does have to be some rethinking about "time since PhD." Time is a complicated issue that needs to be dealt with.

DR. WILLIAMS: The rethinking of time very specifically has to be reflected in the letter that goes out to reviewers. There is an increasing legal dimension to this. In the Lisa Arkin case for example, the $500,000 settlement, one of the statements was that stopping the tenure clock was "a red flag" in a tenure file. This is just not anywhere a university wants to go legally. If you have a stop the tenure clock policy, there should be a strong, concerted effort to make it real, or you are placing yourself at potential risk.

---

[81] S Thornton (2003). Maternity and Childrearing Leave Policies for Faculty: The Legal and Practical Challenges of Complying with Title VII. *University of Southern California Review of Law and Women's Studies* 12(2):161-190.

SECTION 1: SUMMARIES OF CONVOCATION SESSIONS 71

DR. DRAGO: When Joan and I published the part-time tenure track piece,[82] one of the responses we got from a man was, I don't know what half a tenure case is. The time from PhD measure is particularly of concern with more scientists doing postdocs. It is exacerbating the conflict between the biological and tenure clocks. One thing that I keep coming back to is maybe we should be thinking about shortening the tenure clock at some level. Law schools are three years I think.

DR. MARTIN: Law schools are tenure in a flash. I did want to mention one issue that hasn't come up, which is the structure of grants in the sciences. If you tie eligibility for certain grants for three or five or ten years from PhD, that means that if you take time out to have a baby, or if you take time out to care for an elderly person, you are basically placed at a systematic disadvantage. I understand that this is not an easy system to solve, but it is a system that clearly has a disproportionate negative impact on women.

DR. CHAN: Emily Chan, Colorado College. A lot of the data in the points represented so far is about young researchers in R-1 universities, maybe R-2 universities. Given that liberal arts colleges are more and more emphasizing research in the sciences, I wonder how a lot of these things that you talk about may be similar or different in the liberal arts context.

DR. DRAGO: From our study, the main differences were women in liberal arts colleges reported missing more of their children's important events when they were young, because of a heavier teaching load. There is no surprise there. They were more likely to parent. The women and men in chemistry were more likely to parent than the women and men in English. Presumably, that has to do with family. There is about an $8,000 pay difference.

DR. WILLIAMS: There are also different design issues with regard to a part-time tenure track in a liberal arts college.[83]

DR. GINTHER: My research combined all people in four-year institutions. Women tend to be more represented at teaching institutions than at Research 1 institutions.[84]

---

[82]R Drago and JC Williams (2000). A half-time tenure track proposal. *Change* 32(6):46-51.

[83]JC Williams (2004). Part-timers on the tenure track. *The Chronicle of Higher Education,* http://chronicle.com/jobs/2004/10/2004101401c.htm.

[84]DA Nelson and DC Rogers (2005) *A National Analysis of Diversity in Science and Engineering Faculties at Research Universities*, http://www.cwru.edu/admin/aces/search/diversityreport.pdf.

PANEL 4
IMPLEMENTING POLICIES

Panel Summary

Recruitment Practices
Angelica Stacy, Department of Chemistry, University of California, Berkeley

Reaching into Minority Populations
Joan Reede, Harvard Medical School

Creating an Inclusive Work Environment
Sue Rosser, Ivan Allen College, Georgia Institute of Technology

Successful Practices in Industry
Kellee Noonan, Technical Career Path, Hewlett Packard

Selections from the Question and Answer Session
Moderated by committee member ***Nan Keohane***

PANEL SUMMARY

This panel discussed specific practices and policies that foster or discourage the employment and advancement of women in academic science. Angelica Stacy began by questioning whether the entryway to science positions in research universities is inviting to women. Some science departments have noted an "applicant-pool problem" in which the proportion of female applications is lower than that within the total pool of doctorate holders. Departments that have excellent records in hiring women have taken such special steps as selecting diverse search committees and including input from graduate students in the search process.

Like Xie and Drago, Stacy noted that conflict between work and family is a barrier to women, citing statistics that married men with young children are 50% more likely to enter tenure-track jobs than comparable women. Three-fourths of female assistant professors at the University of California, Berkeley have no children, as opposed to 58% of men; and only 9% of female assistant professors have two children, as opposed to 13% of males. Narrow position specifications also disadvantage women, who are 50% more likely than men to do interdisciplinary work and have joint appointments. Berkeley's new department of bioengineering, for example, is 50% female. Building an entry that is inviting and accessible to women requires proactive recruitment, family-friendly policies, and full-time allocations for multidisciplinary posts.

Joan Reede discussed the special problems of women biomedical scientists who are members of minority groups. Their underrepresentation results from a variety of pipeline issues and barriers, despite the fact that interest in studying science is higher among African American and Asian girls than among white girls. Of African Americans who receive science degrees, 64% are women.

Minority women who do enter academic science careers suffer a double jeopardy, however, because of isolation, lack of mentoring, and the expectation that they will serve as advisors and committee members, Reede added. They have only limited networks for their own guidance. She then described examples of several Harvard University programs (aimed at encouraging and supporting minority science and medical students) that have contributed to the more than doubling of underrepresented minority representation on the Harvard Medical School faculty as well as providing increased opportunities to more than 4800 students.

Sue Rosser examined various approaches to creating inclusive work environments. She used questionnaires and interviews to identify the issues of greatest importance to female science faculty. The top issue, cited by 65 to 88% of respondents, was balancing work with family responsibilities. Next were time management, especially the balance among teaching, research, and committee responsibilities; the low number of women and resulting lack of mentoring and camaraderie; the difficulty of gaining respect and credibility among male peers;

and dual career problems, especially pertinent to women scientists since more than 60% are married to men scientists. Reports of harassment, overt discrimination, stereotyping, lack of respect for one's work, and mere "lip service" to diversity were numerous.

The National Science Foundation's ADVANCE program is focused on institutional transformation to facilitate the advancement of women scientists into senior faculty and leadership positions in universities. The 19 universities that have received ADVANCE grants are developing an array of models for transformation. Helpful models for making workplaces more inclusive include family-friendly policies and practices and training search committees, chairs, deans, and tenure and promotion committees.

Kellee Noonan, the only convocation speaker from industry, described her company's Technical Career Program, which is used throughout the company's worldwide operations. The program aims to recruit and advance the careers of a diverse workforce of highly trained technical professionals. It is designed to break the glass ceiling by making processes fair and transparent, eliminating cumulative bias in selection and promotion, and applying to career advancement the First Law of Diversity: "when bad things happen, they happen worst to people in the minority."

The career ladder for each position is clearly open and defined, and promotion is based on criteria that are readily accessible on company Web sites and linked to open and broadly available learning resources. This allows employees to know what they need to do to meet each criterion and then to gain the skills required for advancement. A core team works continuously to educate employees about the criteria, using regular forums and other means. A diversity team focuses on goals and metrics, developing mentoring and other programs to help underrepresented ethnic, gender and geographic groups to succeed and advance. Those policies have helped break the glass ceiling for women technologists in some areas of the company, and current efforts continue to "raise the roof."

RECRUITMENT PRACTICES

Angelica Stacy
Department of Chemistry, University of California, Berkeley

Angelica Stacy focused on recruitment practices, specifically, the structures that are in place in academic institutions and the degree to which they are inviting and accessible to women. As associate vice provost for faculty equity at the University of California, Berkeley, and professor of chemistry, she has been monitoring for a number of years the searches for and the career advancement of faculty and also has been doing general studies of diversity and inclusion on the faculty.

She provided evidence from the Berkeley database on the faculty applicant pool that the entryway to faculty positions is neither inviting nor accessible. Of people who have applied to Berkeley for a faculty position, 75% have filled out a survey indicating their gender and ethnicity. No faculty search had a pool of women that was equal to, let alone above, the pool of women that are in even the most conservative estimate of the PhD pool. Figure 1-10 shows the U.S. PhD pool weighted average across physical science, mathematics, and engineering (n = 12,214). It includes about 15% white women, and 4-6% female members of underrepresented minority groups. Above that is shown the 2001-2004 Berkeley applicant pool (n = 3,952), which is about one-third of the PhD pool. Women are hired as assistant professors in a proportion similar to their representation in the applicant pool.

In the biological and health sciences, there are many more women in the PhD pool (Figure 1-11). The applicant pool is about 17% of the PhD pool. Berkeley has been hiring about 50% women into assistant professor positions and now has close to 50% women at the associate professor level. That bulge has not reached the full-professor or leadership ranks at Berkeley.

What can we do to improve recruitment and hiring of women? Figure 1-12 plots the percentage of women hired against the percentage of women in a conservative estimate of the pool,[85] which the departments estimated themselves. The diamonds represent individual departments at Berkeley. The dashed line is a theoretical indicator for hiring equaling the pool. As shown by the bold line, over the entire institution, and even within the sciences, hiring vs. the pool is on the average about even.

Those data can be used to determine which recruitment and hiring practices correlate with hiring above, at, or below the applicant pool. Table 1-1 shows a ranked list of practices used by departments to enhance the faculty pool; 96% of departments reported listing faculty positions in multiple venues and 84% said that they made it clear that women and members of underrepresented minority groups were encouraged to apply. Dividing the list by those departments that hired at or above the level of the pool of women (Exc.), and those that hired below the level of the pool (Not Ex.) yielded statistically significant differences. The most significant ones—designating an affirmative action officer to serve on the search and saying women and minorities please apply—were highly correlated with those departments that didn't hire at the level of the pool. Those who are doing excellently are using other kinds of strategies: they are including graduate student input, selecting diverse search committees, and going out to professional meetings and establishing relationships and inviting women to apply

---

[85]The pool was calculated on the basis of on PhDs granted to US residents, 1997-2001 (Survey of Earned Doctorates, National Science Foundation) at the 35 top-quartile rated doctoral programs (National Research Council reputation ratings) producing the most PhDs.

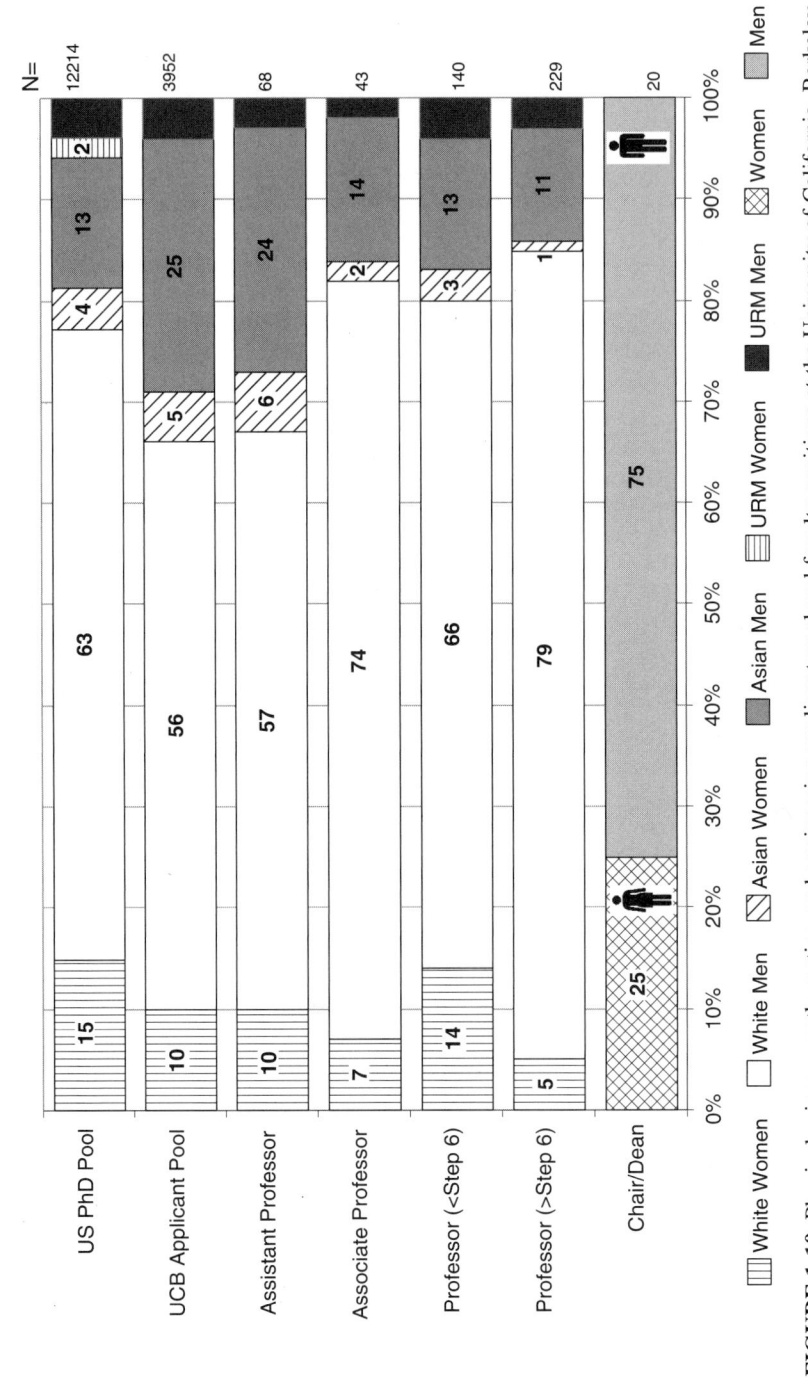

FIGURE 1-10 Physical science, mathematics, and engineering applicant pool and faculty positions at the University of California, Berkeley.
SOURCE: UC Berkeley Faculty Applicant Pool Database, 2001-2004; UC Berkeley Faculty Personnel Records, 2003.

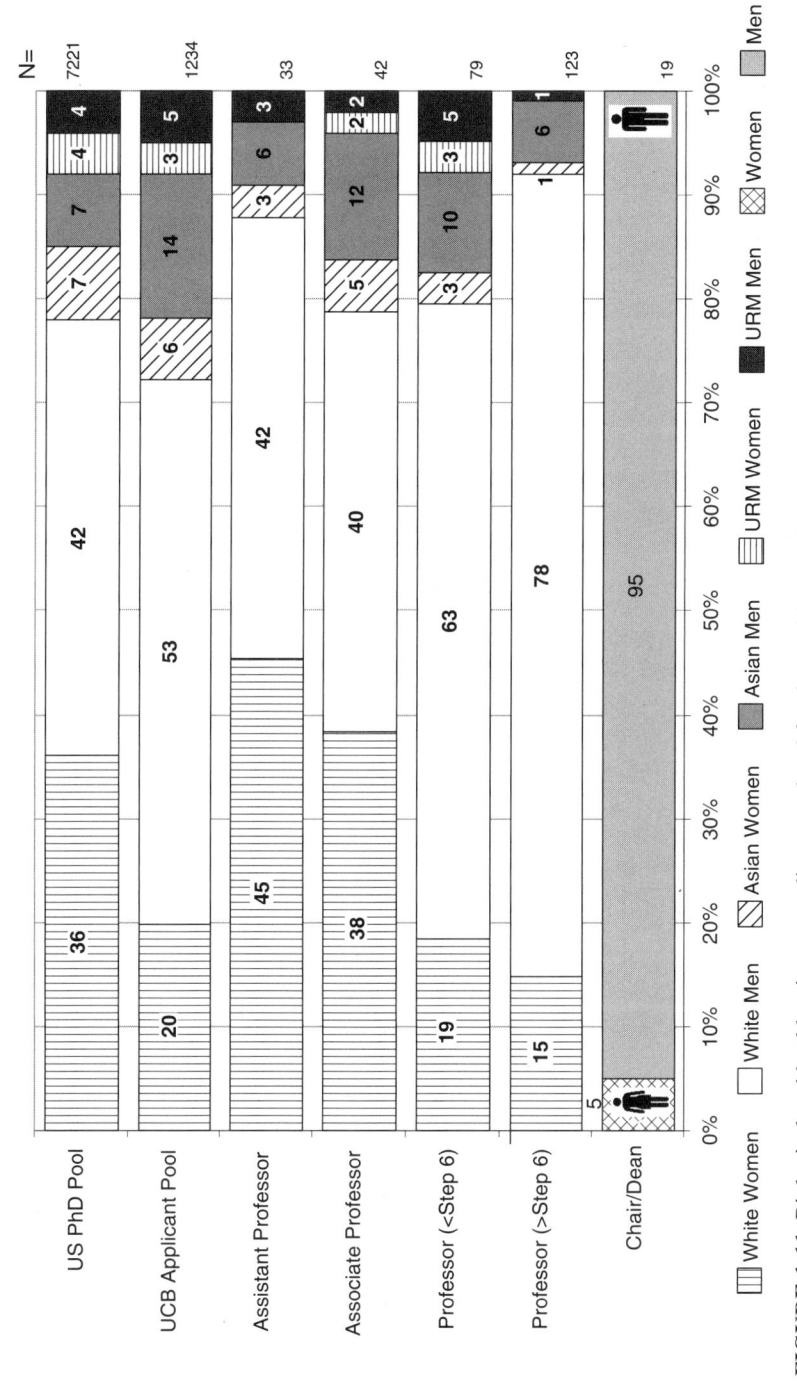

FIGURE 1-11 Biological and health sciences applicant pool and faculty positions at the University of California, Berkeley.
SOURCE: UC Berkeley Faculty Applicant Pool Database, 2001-2004; UC Berkeley Faculty Personnel Records, 2003.

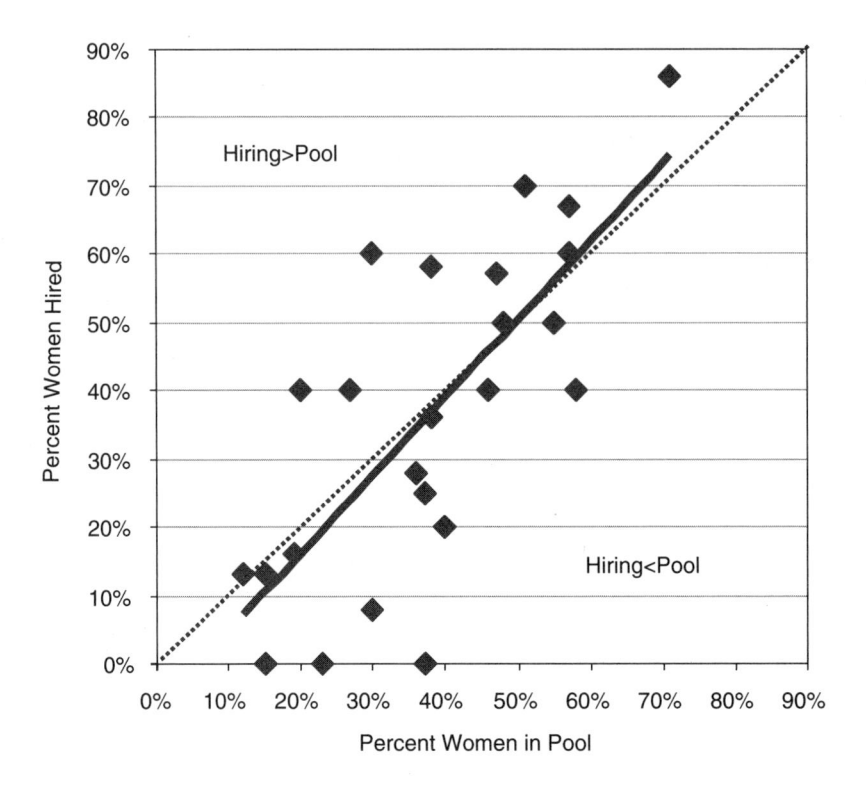

78   COMPONENTS OF SUCCESS FOR WOMEN IN ACADEMIC SCIENCE & ENGINEERING

FIGURE 1-12 Departmental hiring vs the applicant pool, University of California, Berkeley.
Notes: Figures are since 2000; only departments that hired over five faculty during that period are included.

rather than assuming that women feel confident enough or included enough to send in an application.

Work-family conflict also affects the applicant pool. Mary Ann Mason and Marc Goulden have found that married women who have children pay a 50% penalty in terms of entering faculty positions, as compared with single women or married men who have children.[86] At Berkeley, of female assistant professors, 16% have one child and 75% have no children; 27% of male assistant professors have one child, and only 58% have no children (Figure 1-13).

---

[86]M Mason and M Goulden (2004). Marriage and baby blue: Redefining gender equity in the academy. *The Annals of the American Academy of Political Social Science* 596:86-103.

TABLE 1-1 Methods Used by University of California, Berkeley Departments to Enhance Faculty Hiring Pool

Rank Order	Methods Used	# Women Hired Exc. (n=25)	Not Ex. (n=29)	Total (n=59)
1	Listed faculty positions in multiple venues	96%	97%	96%
2	Job description made clear women/URM encouraged to apply	76%	90%	84%
3	Made personal calls to encourage potential candidates to apply	84%	86%	84%
4	Selected diverse search committees	92%	79%	84%
5	Included graduate student input in search process	92%	72%	82%
6	Made calls to colleagues asking them to encourage women/URM to apply	80%	83%	80%
7	Circulated job description among networks women/URM educators	88%	72%	79%
8	Designated an affirmative action officer to serve on search	64%	90%	77%
9	Approached or interviewed applicants at professional meetings	72%	72%	73%
10	Established relationships with local/national women/URM organizations	68%	52%	59%
11	Educated search committee members on diversity/equity/affirmative action	52%	55%	54%
12	Discounted caregiving related resume gaps	32%	41%	36%
13	Prioritized subdisciplines with high diversity	36%	31%	32%
14	Encouraged UC President's Postdoctoral Fellows to apply	36%	31%	32%
15	Interviewed candidates at a variety of conferences	36%	21%	27%

Notes: Hatched shading denotes $p < .05$ significant difference based on chi-square. Dotted shading denotes $p < .10$ significant difference based on chi-square.

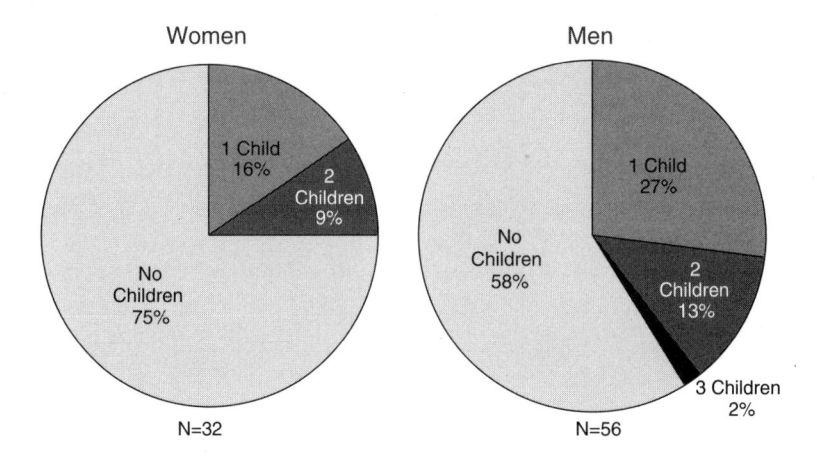

FIGURE 1-13 Children in households among assistant professors at the University of California, Berkeley.
SOURCE: MA Mason, A Stacy, and M Goulden. 2003. "The UC Faculty Work and Family Survey." See http://ucfamilyedge.berkeley.edu.
Note: Numbers of children were self-reported.

> *We now have better policies around the country. Let me assure you that if you get out there and start to say, "This is what we want, this is the way we do things, and this is an entitlement," which is where a number of our institutions are moving, I think we're going to see things change.*
>
> —*Angelica Stacy*

Narrow position specifications also affect the applicant pool and numbers of women hired. There is mounting evidence that women are choosing to work at the boundaries of the disciplines. Among the STEM[87] faculty at Berkeley, 26% of the women and 15% of the men have joint appointments. Women tend to hold joint appointments in business, biology, law, city and regional planning, economics, and environmental science. In one of the newer departments, bioengineering, 50% of the faculty are women. The biological sciences were restructured. They now include broad, multidisciplinary approaches, and no longer have the old embedded departmental structures of the beginning of the last century. Fifty percent of the faculty are women.

---

[87]At Berkeley, biology and health sciences are not included in the category "STEM", science, technology, engineering, and mathematics.

> *I can't tell you how many times I have reviewed searches in which the people—predominantly women and minority-group members—were not hired, because they didn't "fit."*
>
> —*Angelica Stacy*

As part of its diversity initiative, Berkeley has started to hold full-time equivalent (FTE) faculty positions centrally for groups of faculty and departments that get together proposing new multidisciplinary research areas. This is done to counteract the tendency in departments to hire people who they have always hired, who look just like them, who fill the mainstream slots, rather than moving the institutions forward into new areas. For this, institutional leadership is important. Stacy concluded with four main ideas: proactive recruitment, family-friendly policies, FTE allocations, and leadership. Her motto: build it, so the best will come.

REACHING INTO MINORITY POPULATIONS

Joan Reede
Harvard Medical School

Joan Reede focused her remarks on reaching into and across minority populations, particularly in the biomedical sciences.

There is a well-known persistent and continuing underrepresentation of African American, Hispanic, American Indian, Alaskan Native, and Native Hawaiian males and females in academic science. The underrepresentation is fueled by limitations in the pipeline and by what Reede termed the "academic black hole" into which many graduates fall, a hole associated with attrition and lack of advancement. Pipeline deficiencies are found in access, achievement, and attitude. Oftentimes, minority-group students are faced with inadequate preparation and awareness of opportunities, underdeveloped relationships with adults and an associated unrecognized potential, and lack of mentoring and career counseling. Minority-group students often have insufficient social supports and resources, particularly financial resources that are necessary to pursue advanced education. As described by Toni Schmader earlier, data show that African American, Hispanic and American Indian students fare less well on high school, college and professional school standardized tests.

In an analysis of the National Educational Longitudinal Survey, Hanson found that there was variability in attitudes toward science for women across racial and ethnic groups.[88] For example, African American female students

---

[88]SL Hanson (2004). African American women in science: Experiences from high school through the post-secondary years and beyond. *NWSA Journal* 16(1):96-115.

expressed a greater interest in science than did white female students in the 8th and 10th grades.

> *An important question is what factors sustain, increase, or decrease student's interests as they move along the educational ladder. Once minority-group members and women enter the academy, they are confronted with barriers and diversity taxes—such barriers as assumptions and stereotypes, including all the "isms"—racism, classism, sexism—and discrimination. They often face cultural, social, and intellectual isolation, and do not have access to formal and informal mentoring.*
> —*Joan Reede*

Minority-group women face a double jeopardy associated with their limited numbers and are expected to take on extra service responsibilities associated with counseling of students, residents and fellows and to assume committee assignments. Those activities are not adequately acknowledged within the academic reward system or in the promotion review process.

Minority groups and women often have few research sponsors and opportunities for collaborative research. Their informational networks that can provide input, critique, validation of experiences, and an understanding of organizational rules and bureaucracy are limited. Those issues are cumulative and persist from junior to senior faculty levels.

Adequate numbers of minority-group members, particularly female faculty, cannot be achieved unless pipeline issues are dealt with. Deficiencies in the educational process leading to graduate and professional school disproportionately affect minority-group and poor students. An analysis by the Education Trust[89] found that of every 100 white kindergartners, 93% would graduate from high school, 65% would complete some college, and 33% would obtain bachelor's degrees. The corresponding numbers for black kindergartners were 87%, 50%, and 18% respectively. For Latino and American Indian kindergartners, only 11% and 7% of youth, respectively, would earn bachelor's degrees. There is also an association between poverty and graduation: the vast majority of students who graduated from college by the age of 26 years come from high-income families.

Of those students who enter college, the National Science Foundation reports that the percentage of Asian, African American, and Latino freshmen who intended to pursue a science or engineering major is higher than that of white freshmen. And for all racial and ethnic groups, the percentage of freshmen females planning to major in science or engineering was higher than the percentage of males. That was true for all science and engineering majors (Table 1-2).

---

[89]Education Trust, Inc. (2002). US Department of Commerce, Bureau of the Census, March Current Population Surveys, 1971-2001. In *The Condition of Education*. US Department of Education.

TABLE 1-2 Intentions of Freshman to Major in Science and Engineering Fields, by Race, Ethnicity, and Sex, 2002

Race/Ethnicity	All S&E Majors (%)	Biological/ Agricultural Sciences (%)
White		
Male	23.8	6.2
Female	*37.9*	*7.6*
Asian/Pacific Islander		
Male	33.1	10.2
Female	*53.0*	*13.5*
African-American/Black		
Male	31.9	5.8
Female	*38.0*	*10.0*
Chicano/Puerto Rican		
Male	30.6	6.8
Female	*39.4*	*9.2*
Other Hispanic		
Male	30.9	6.9
Female	*40.8*	*8.3*
American Indian/Alaska Native		
Male	27.2	6.1
Female	*37.6*	*8.8*

SOURCE: Women, Minorities, and Persons with Disabilities in Science and Engineering, 2004, National Science Foundation.

From 1994 to 2001, there was an increase of 27 to 38% in the numbers of science and engineering bachelor's degrees awarded to all minorities (Figure 1-14). During that period, however, there was a 10% decline in science and engineering bachelor's degrees awarded to whites. Much of the increase among minorities was fueled by an increase in science and engineering degrees awarded to women. For example, in 2001, 64% or roughly 21,000 of science and engineering bachelor's degrees earned by African Americans, and 55% or 15,000 of the science and engineering bachelor's degrees earned by Hispanics were awarded to women.

There is a similar increase in science and engineering doctorates awarded to minority women in the same period, except for Asian Americans (Figure 1-15). Although there was an increase in absolute numbers, the representation of minority-group women as a percentage of all science and engineering doctorates in 2001 was less than 9%, and half of those degrees were to Asian American women. The decrease in Asian American women receiving science and engineering doctoral degrees was not seen in the biological sciences, where the numbers were the same in 1994 and 2001: 268 degrees.

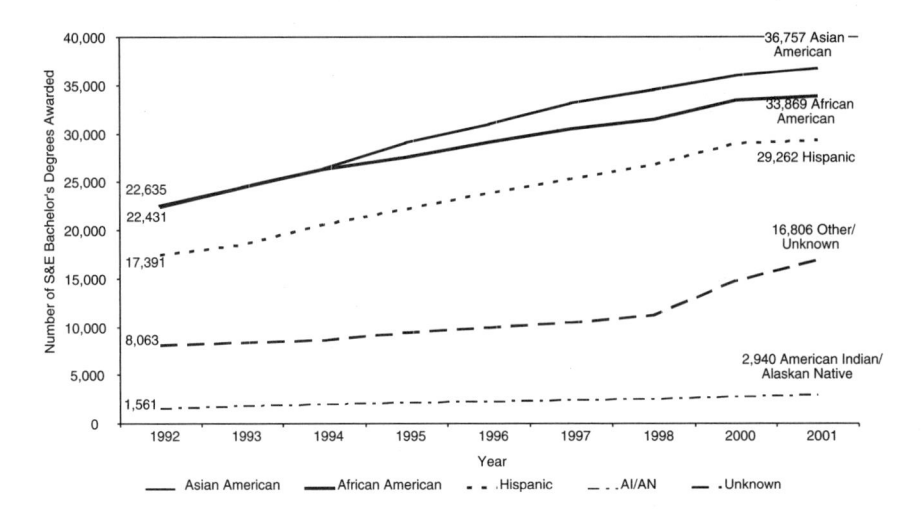

84   COMPONENTS OF SUCCESS FOR WOMEN IN ACADEMIC SCIENCE & ENGINEERING

FIGURE 1-14 Number of science and engineering bachelor's degrees awarded to minority females, by race and ethnicity, 1994-2001.
Note: *American Indian/Alaskan Native* includes Native Hawaiians and other Pacific Islanders; *Other/Unknown* includes those with unknown race/ethnicity and respondents choosing multiple races (excluding those selecting Hispanic ethnicity).
SOURCE: National Science Foundation, Division of Science Resources Statistics, special tabulations of US Department of Education, National Center for Education Statistics, Integrated Postsecondary Education Data System, Completions Survey, 2001.

For academic medicine and the pipeline of medical students, there was an overall decline in medical-school applications from all racial and ethnic groups from 1997 to 2002. That was followed by a rise in the past two years, with underrepresented-minority applicants finally achieving their 1992 levels in 2004. Associated with the overall increase in applications was an increase in applications from women. However, there was variability in applications across racial and ethnic groups; African American women were nearly 70% of all African American applicants to medical schools.

In looking at applications, matriculants, and graduates, it is important to disaggregate racial and ethnic groups. For example, there was variation in applications among Hispanic subgroups, with a 10% increase in Mexican American applicants from 2002 to 2004, and a 20% decline in Puerto Rican applicants in the same period. That variability was also seen in the percentage of women among the various racial and ethnic applicant pools.

Among medical school faculty, three striking patterns are noted: men and women of color are underrepresented; African American, Hispanic, American Indian, and Alaskan Native women represent a very small percentage of all

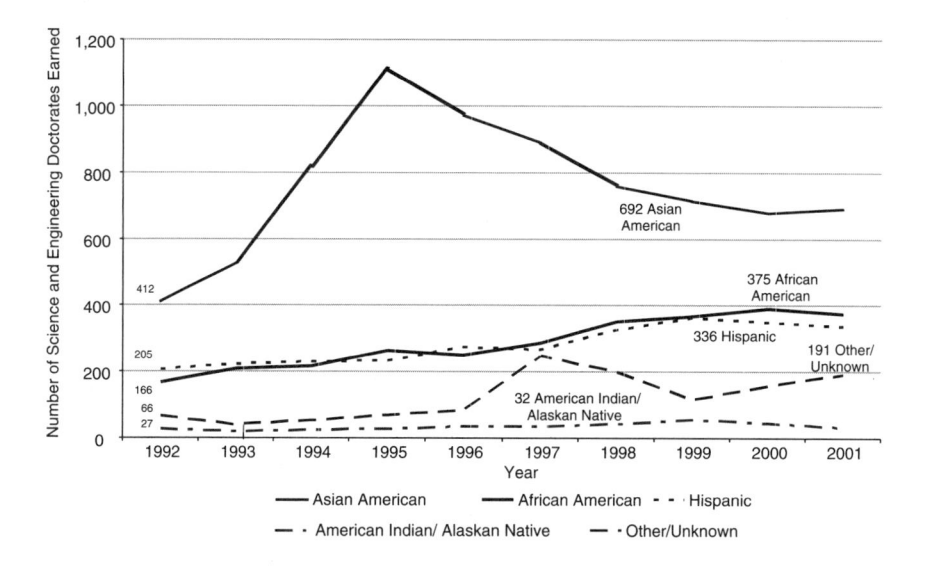

FIGURE 1-15 Number of science and engineering doctorates awarded to minority-group women, by race and ethnicity, 1994-2001.
Note: *American Indian/Alaskan Native* includes Native Hawaiians and other Pacific Islanders; *Other/Unknown* includes those with unknown race/ethnicity and respondents choosing multiple races (excluding those selecting Hispanic ethnicity).
SOURCE: National Science Foundation, Division of Science Resources Statistics, Survey of Earned Doctorates, 1994-2001.

medical school faculty; and the proportion of women faculty in all racial and ethnic categories declines in advancing up the academic ladder from instructor to full professor (Figure 1-16).

Among science and engineering doctorate holders employed in colleges and universities, similar patters of underrepresentation of racial and ethnic minorities, low numbers of women faculty, and aggregation of women among lower academic ranks are also seen (Figure 1-17).

Part of Harvard Medical School's response to the need for diversity was the establishment of the Minority Faculty Development Program in 1990 and its incorporation into the Office of Diversity and Community Partnership (DCP) that was established in 2002. DCP sponsors almost 20 programs that cross multiple academic levels from kindergarten through college and medical student fellowship and junior faculty programs that provide multiple points of entry, exit, and re-entry. Themes included in the development and implementation of DCP programs include continuity, collaboration and partnership, the building of networks and support systems, formal and informal mentoring, skill building, increasing awareness of career paths and opportunities, and evaluation and tracking.

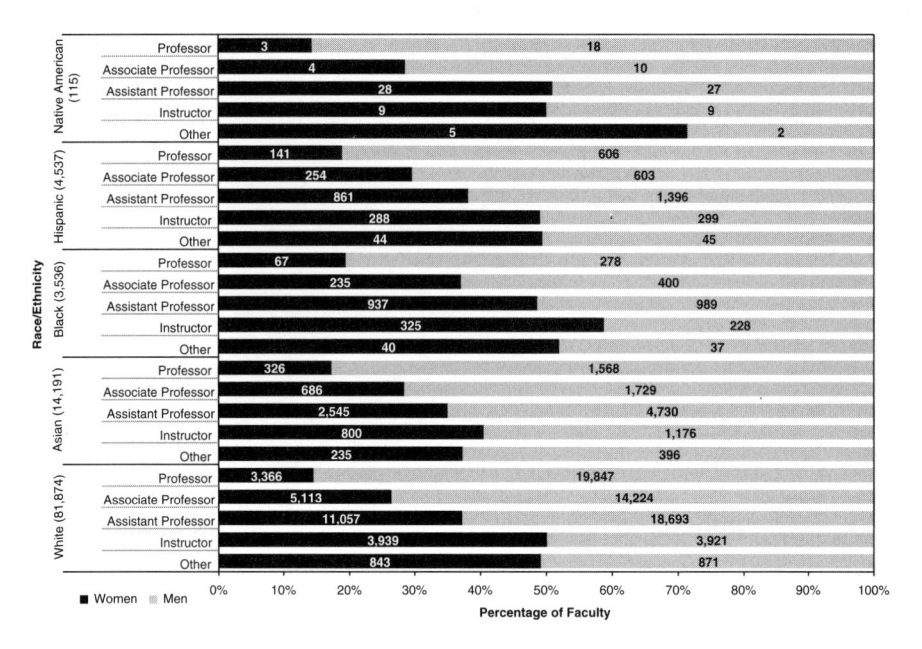

86 COMPONENTS OF SUCCESS FOR WOMEN IN ACADEMIC SCIENCE & ENGINEERING

FIGURE 1-16 Medical school faculty by rank, gender, race, and ethnicity.
SOURCE: AAMC Faculty Roles Survey, 2004.

Using elements of those themes, Reede described three Harvard Medical School programs that are related to organizational structures and policies that deal with collaboration and that address issues that cross multiple levels of the academic and career ladder.

Biomedical Science Careers Program (BSCP). Of the students in this program 60% are women, 47% are African American, and 19% are Hispanic. Founded in 1991 with a group of people from Reede's office at Harvard Medical School, the Massachusetts Medical Society, and the New England Board of Higher Education, the BSCP quickly grew to a community of individuals and organizations that shared a desire to address diversity. It is now led by a board of directors that includes presidents and CEOs in biotechnology, medical device research, legal, and finance industries; leaders in academe, professional associations, and community colleges; educators; practitioners; and employers. Supported by the community, and without public funding, the BSCP has now reached more than 4,800 high school, college, medical school and professional school graduates, and postdoctoral students and fellows. The more than 500 volunteers who have made the BSCP initiatives work point to the fact that many in the biomedical community are deeply concerned about education, diversity, and the

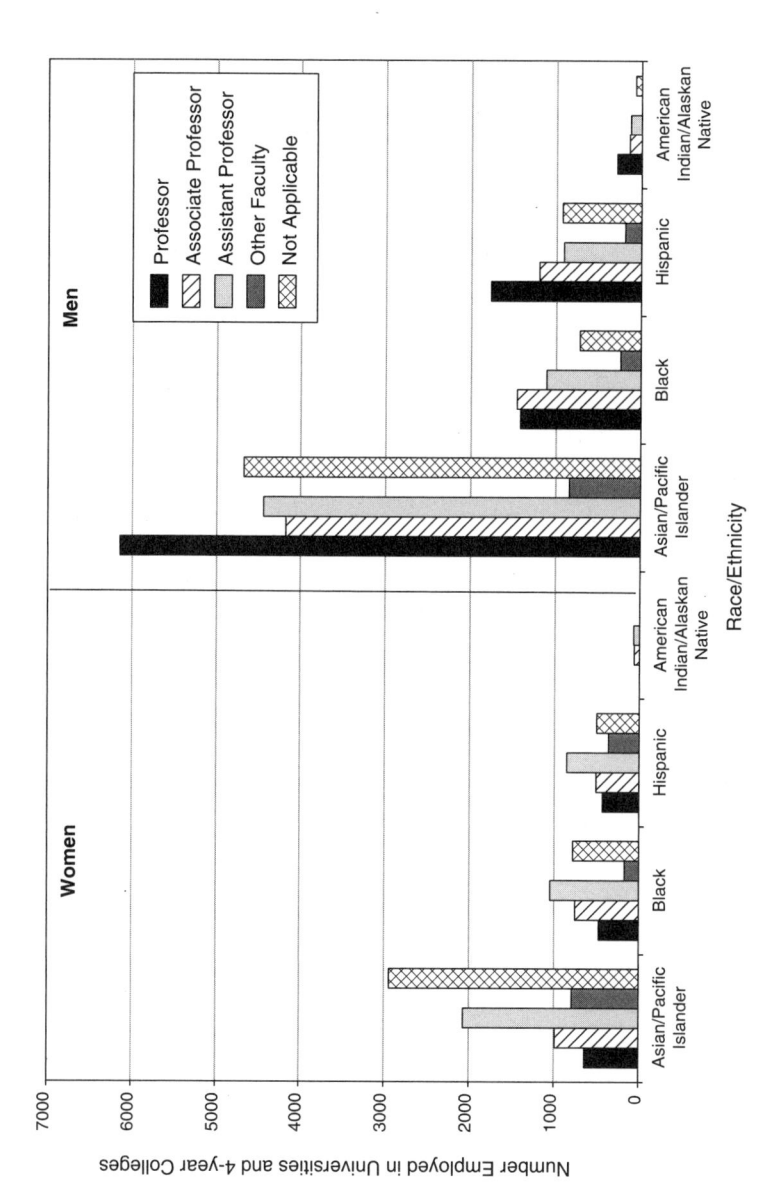

FIGURE 1-17 Number of science and engineering doctorate holders employed in science and engineering occupations in universities and 4-year colleges, by race, ethnicity, and faculty rank, 2001.
SOURCE: Women, Minorities, and Persons with Disabilities in Science and Engineering (2004). Arlington, VA: National Science Foundation.
Note: American Indian/Alaskan Native (AI/AN) data for "Other faculty" are suppressed because there are fewer than 50 weighted cases.

future science workforce. BSCP students have said that experiential opportunities such as internships and job shadowing, contact with minority-group role models, and encouragement from teachers have a large influence on their educational and career goals. BSCP has influenced students to obtain more information, strengthen their interests, and to make them aware of career opportunities and connections with people. Students have been able to identify jobs and apply for jobs, participate in new programs and internships, identify mentors at their schools, and obtain funding.

Visiting Clerkship Program (VCP). More than 700 third- and fourth-year medical students from schools across the country have participated in this 1-month externship program, established in 1990. The program offers travel, housing, faculty advisers, and access to networks and resources. Some 15% of VCP students have returned to Harvard as residents, fellows, and faculty. The students have said the things that are important to them in selecting a residency program are academic training programs, the pre-eminence of those programs, their recommendations and interactions with advisers and mentors, their potential for research participation, and family considerations.

Center of Excellence in Minority Health and Health Disparities. This fellowship program for junior faculty was established in 2002; the first cohort began in 2003. To date, nine faculty fellows have participated. Four have been promoted, one to a division chief. Two are up for promotion now. Eight have obtained external grant funding. An essential component of the program is selected mentors. All fellows have to have letters of support and involvement of department chairs, and the presidents of the participating Harvard Medical School hospitals are involved in the selection. Built into this program are accountability and recognition of and support for excellence.

> **What issues need to be addressed if we are to achieve racial and ethnic diversity? Responsibility at multiple levels. There needs to be top-down and bottom-up activity that provides vehicles to ensure the success of minority and women students, trainees, and faculty. And this activity must extend beyond verbal acknowledgment of the need for diversity. Recognition of the need should be incorporated into the institutional missions, reiterated in the setting of policies, and integrated in the design of programs. Data should be disaggregated to ensure that issues that disproportionately impact certain racial and ethnic groups are appropriately addressed and outcomes of interventions are adequately tracked. Diversity is not just about minorities and women. Diversity is about how we can improve and advance science for all.**
> —*Joan Reede*

CREATING AN INCLUSIVE WORK ENVIRONMENT[90]

Sue Rosser
Ivan Allen College, Georgia Institute of Technology

Sue Rosser began with an emphasis on the need for practical institutional approaches, as suggested by the MIT report.[91] Almost simultaneous with the release of the MIT report, the National Science Foundation launched its ADVANCE Institutional Transformation Initiative. For many years, there had been successor programs such as visiting professorships for women, career advancement awards, and professional opportunities for women in research and education (POWRE). Although some of them had a component for institutional transformation, they largely gave money to individual women researchers. In contrast, ADVANCE focuses on institutional changes, especially for women on the academic tenure track to senior and leadership positions. The first round of ADVANCE awardees occurred in 2001 and the second in 2003, and the third round will be announced in early 2006.[92] From these grants will come several models of what has worked and what has not worked for different institutions.

Rosser studied the NSF POWRE awardees and the Clare Booth Luce professors to understand the most significant issues, challenges, and opportunities facing women scientists today as they plan their careers.[93] She received about 450 responses to e-mail questionnaires and conducted 40 in-depth interviews. Respondents were distributed among all the disciplines, and each of the different years the awards were made are represented.

> *The first question—an open-ended question—was, What are the most significant issues, challenges, and opportunities facing women scientists today as they plan their careers? People could have said anything. What amazed me was that balancing career with family was the overwhelming response—65 to 88% for all 4 years.*
>
> —Sue Rosser

After balancing career with family, a second major issue was time management: balancing work with research, teaching, and service. The third issue was isolation: low numbers, and lack of camaraderie and mentoring. The fourth issue

---

[90]For more details, figures, and references, see the paper by Rosser in Section 2.

[91]Massachusetts Institute of Technology (1999). A study on the status of women faculty in science at MIT. *The MIT Faculty Newsletter* 11(4):14-26, http://web.mit.edu/fnl/women/women.html.

[92]Several of the meeting posters presented research from ADVANCE grantees; see the poster abstracts later in this volume (p. 175).

[93]SV Rosser and J Daniels (2004). Widening paths to success, improving the environment, and moving toward lessons learned from experiences of POWRE and CBL awardees. *Journal of Women and Minorities in Science and Engineering* 10(2):131-148.

was gaining credibility and respectability. And then the fifth major issue was the dual career problem. All those issues have come up previously in this meeting.

> *Many of these issues are centered around the fact that the life cycle is based on what I call the white male model. There is nothing wrong with that, unless you're not white and male; if you are not then it does not work very well for you, particularly the competition between the biological clock and the tenure clock.*
>
> —*Sue Rosser*

Another way of presenting the data is by dividing the responses into four groups:

1. Pressures women face in balancing career and family (~30%).
2. Problems faced by women because of low numbers and stereotypes held by others regarding gender (~10%).
3. Issues that are faced by both men and women scientists and engineers which, because of the current environment of tight resources, may pose particular difficulties for women (~7%).
4. Overt discrimination and harassment (~5%).

Included in the first category are issues related to the dual career family, which is a particular problem because most women scientists and engineers are married to men scientists and engineers.

The second category has to do with being taken seriously and having increased visibility. The latter is particularly important for women of color, who are very visible because of their low numbers in faculties. If things go well, that's remembered and can put you on a quick career trajectory. If things go badly, it is not forgotten.

The third category contains issues that are faced by everyone but that have particular angles for women, such as the assumption of being available.

Finally there are overt discrimination and harassment, including slow promotions, lack of women in senior positions, and placement of women into difficult situations because they must buffer the bad behavior of their male colleagues.

Effective models for countering some of those issues have been developed and tested at some of the ADVANCE institutions, including

- Family-friendly policies and practices, including family-work initiatives, tenure-clock extension, childbearing and family leave, active service modified duty, and daycare facilities.
- Training of search committees.
- Training of chairs and deans to manage search committee results and to foster a welcoming departmental environment.
- Speed mentoring.

Rosser explained that the Georgia Institute of Technology, an ADVANCE institution, has focused on training of tenure and promotion committees. It developed an ADEPT model, an interactive CD-ROM with which people can participate in a tenure and promotion meeting. It developed nine case studies and nine virtual CVs to go with them. The player of the game can participate with three virtual people in the tenure meeting; and depending on what the player says, ADEPT sends the conversation in a particular direction. In addition to research expertise, ADEPT includes such issues as gender, disability, race, ethnicity, and sexuality. The deans were the first to use ADEPT, and then it moved to the department level, the chairs, and all the promotion and tenure committees. All faculty are now using ADEPT.

Rosser and her colleagues are now developing a "navigate your career" section for junior faculty, with frequently asked questions, such as, Should I serve on that NSF review panel, or should I be writing my own proposal? How do I decline gracefully to serve on that committee that my senior colleague has asked me to serve on?

"Speed mentoring" is another popular program at the Georgia Institute of Technology. Junior faculty take their CVs to a meeting with senior faculty who have served on tenure and promotion committees, but who are not currently on tenure and promotion committees. The junior faculty meet with four or five senior faculty in an hour to get a quick take on their CVs and what they might need to do to get ready for promotion to the next level. The senior faculty may suggest another publication or two in a refereed journal, beefing up their teaching, more service on national committees, and so on. Junior faculty like this very much, and say that they get an impression of the different perspectives that different people have. These are the people who have served on tenure and promotion committees, so it is quite realistic.

Rosser is doing some research on older women scientists because she has become concerned that many of the policies put into place through ADVANCE are primarily for younger women. It is very important for more junior women to achieve tenure, but there are different problems for senior women that need to be addressed.

SUCCESSFUL PRACTICES IN INDUSTRY

Kellee Noonan
Diversity Program Manager, Technical Career Path, Hewlett Packard

Kellee Noonan explained that about 5 years ago, when Hewlett Packard (HP) merged with Compaq, they found that they had noncompatible technical career ladders. They took the opportunity to step back and say, "Can we take the former HP program and the former Compaq program, do some industry benchmarking, and put together a framework that gives us a place for technologists to be able to see where they are in terms of their skills and abilities, and what they need to be able to do to move to the next level?"

In creating the Technical Career Program (TCP), the goals were to make the promotion process fair and transparent and to eliminate the cumulative bias in selection and promoting. The TCP appears to be helping in terms of moving women and, in the United States, members of underrepresented groups up the career ladder.

> *The first law of diversity is that when bad things happen, they happen worse to people who are not in the majority.*
> —*Kellee Noonan, quoting Alan Fisher, iCarnegie,*
> *Co-author of "Unlocking the Clubhouse"*

In developing a promotion policy, HP is faced with some challenges different from universities. For example, because it is a global company, the framework needs to apply to all its businesses in the whole world. HP has research laboratories in the United States, Israel, the United Kingdom, China, Brazil, and India. And it has researchers in outposts in many other places. The program has to be able to address all those areas.

The TCP has clearly defined steps, and they are criteria-based. An employee can go to the TCP Web site in the company portal, and ask, What are the criteria to get to the next step? What kinds of things do I have to do? The criteria are balanced around three areas: impact on the business, depth and breadth of knowledge, and technical leadership skills. It's a balance of all of them. You don't have to be perfect in every one. You might have a technologist who is an inch deep but a mile wide. Or you could have a technologist who is an inch wide and a mile deep. What's the difference? How do you evaluate that? The difference is in the impact they have on the business. How are they applying what they are doing to move the business forward and to get the best products, the best technology, and the best services out to customers in a way that matters?

In addition to the criteria, which are specific, there are career development road maps, examples at every level of what it means to meet a criterion. An HP team works diligently to go through the company's workforce-development resources to find the exact resources that match a criterion for each level, and publishes them on the company Web site.

> *If my manager told me that I need to increase my ability to influence a negotiation when there is technology involved, I can go to the TCP Portal Web site and look it up under "influencing" or "negotiation" for my level. And I can find classes that are geared toward improving that skill. We have review boards in place at the higher levels of our ladder. If you get to what we call our strategists' level, you are reviewed*

by a cross-business review board to make sure that we maintain consistency in the application of the levels across the company.
—Kellee Noonan

The TCP program core team has a representative of every chief technology office (CTO) to ensure consistency of policy and interpretation. The core team has an information technology function and a research and development function, called the Office of Strategy and Technology. It organizes quarterly TCP information forums, called "airing the dirty linen time," for employees, HR, managers, and review boards. People can get to the really hard questions. This year, one major focus is the worldwide diversity team. Now, diversity goals and metrics are presented to the CTO every quarter.

HP has also instituted technical leadership curriculum[94] which is for distinguished and master level technologists. Those leadership programs have taken traditional leadership out of the management curriculum. These new programs are focused on leading by influence and engagement, because most of the people on the technical career path don't have a direct staff that can work with them or for them. They have to convince people that it's worth working for them, they have to convince their manager, they have to convince their manager's manager.

We have broken the glass ceiling at some of our levels, but we still need to continue raising the roof.
—Kellee Noonan

SELECTIONS FROM THE QUESTION AND ANSWER SESSION

DR. REED: Alyson Reed with the National Post-Doctoral Association. I was hoping that Sue Rosser could elaborate briefly on the speed mentoring model.

DR. ROSSER: This is a neat program that one of our ADVANCE professors, Jane Ammons, invented. In an hour, junior faculty meet with four or five senior faculty—who have served on tenure and promotion committees, but who are not on tenure and promotion committees at the time of the meeting—and get a quick take on their CV and the senior faculty member's concept of what they might need to do to get ready for promotion to the next level, from another publication or two in a refereed journal, to teaching, or service on national committees. It's been very popular. We have done it now three or four times, and we have calls for more. I highly recommend it. The only issue is getting it organized, but we have

---

[94]The curriculum is based in part on Robert Kelley's *Star at Work* (Three Rivers, MI: Three Rivers Press, 1999) and a program developed in-house called TCP Catalyst.

not had trouble getting senior faculty to do it, and the junior faculty really, really like it.

DR. FLETCHER: Hi, my name is Mary Ann Fletcher at the University of Miami. I would like to speak just briefly about the problem with recruitment of faculty. In my opinion, this problem frequently lies in the fact that at many universities, including my own, have a very severe shortage of women in higher positions in the administration. We have few women as deans, few women as chairs of departments. And so, this leads to search committees that I think frequently have an inborn bias.

I serve on the faculty senate, and we recently had a provost search. I was able to convince our senate to set up a search committee that had one-half female faculty members, and one-half males. Some of the senators said, well, the faculty is not that way. And I said, but the student body is. There are more women than men in our student body. It takes work all the time, and I think the search committees are really key.

DR. AGOGINO: Alice Agogino from the University of California at Berkeley. Do you have any recommendations on how to improve the climate for women faculty in those departments that haven't reached this magical 18% number?

DR. STACY: We really do need to start taking action at the department level, especially when we start to hear that they are not being managed well, or the interactions aren't productive, and that they are not an inclusive environment for all the members of that unit, including students and staff, as well as faculty. I think we just need the wherewithal to say that mismanagement is not acceptable at our institutions.

DR. ROSSER: One of the things that's been helpful to us at Georgia Tech with the ADVANCE institutional transformation grant is in each college we have an ADVANCE professor. This senior professor gets paid extra money from the grant, on the order of $60,000 a year—it's like an endowed chair—to do activities and build mentoring networks. That has united women in each college, so that for example in engineering, where there are something like 415 tenure track faculty and where women may be isolated in departments, they now don't feel as isolated, because they now know each other across the college. I have also encouraged department chairs to encourage their women faculty to join women's studies, which most science chairs think, huh? Why would they do that? I say that may be what makes women hang in.

DR. REEDE: I just want to speak a little bit related to that question, and the comment before about representation of women on searches, et cetera, because one of the critical issues here is oftentimes what gets left out of these discussions: minority representation.

What ends up happening is that one minority person in that department gets rotated for everything. My challenge to the committee and to all of you, as you go back to your institutions, as you think about issues of women, don't ignore the issues of women of color. When you think about putting women on committees, also think about who will be attending to issues for women and men of color. If you are waiting until women and men of color reach a critical mass on faculties, then I'll have to look at my great-grandchildren. Oftentimes with minorities, issues are not going to bubble up to the top. They are so isolated and so alienated you may not hear about it.

DR. HAZELTINE: Florence Hazeltine, National Institutes of Health. You said that that there was a bias that women would get short listed or interviewed, but not get past the next step. This reminded me of some business models where if a woman at a high level lost her job, it took her twice as long to get a job as a man. What I want to know is when women do get interviewed for high-level academic jobs—and I see women presidents and chancellors—how many times have they gone up for it versus how many times the successful men have. If the women knew that in advance, we might be able to get them better coaching, a better feel for the system, when they are just going for practice, and when they are just going for real.

DR. KEOHANE: Having had prior experience either in a coaching setting or having been through another interview is very important. I know that this is true for many men as well as women. Since our particular focus here is for women, I think that the suggestion that was implicit in your question is that you prepare in advance for a job interview in the same way you did as a graduate student when you were going for your first assistant professorship, and you learn from these experiences.

I don't know that there is any evidence that women in most of the really top positions are more likely to be short listed and not chosen. I think that was often true in the past. I think women are more often getting into top leadership positions when they are on the final choice set. What I worry about is that not enough women in academia are seeing high-level leadership positions as an appropriate ambition for themselves. One of the things I find most encouraging about this congregation today is the number of women of strong faculty backgrounds who have been willing to say I will be an associate dean, I will be an associate provost, I will be a provost or a dean or a president.

Section 2

Selected Workshop Papers

Donna Ginther
The Economics of Gender Differences in Employment Outcomes in Academia

Diane Halpern
Biopsychosocial Contributions to Cognitive Performance

Janet Shibley Hyde
Women in Science and Mathematics: Gender Similarities in Abilities and Sociocultural Forces

Sue V. Rosser
Creating an Inclusive Work Environment

Joan C. Williams
Long Time No See: Why Are There Still So Few Women in Academic Science and Engineering?

Yu Xie
Social Influences on Science and Engineering Career Decisions

THE ECONOMICS OF GENDER DIFFERENCES IN EMPLOYMENT OUTCOMES IN ACADEMIA*

Donna K. Ginther
Department of Economics
University of Kansas

Abstract

This paper summarizes research that examines the relationship between hiring, promotion, and salary for tenure track science and social science faculty using data from the Survey of Doctorate Recipients (SDR). Gender differences in hiring and promotion can be explained by observable characteristics. However, gender differences in salaries persist at the full professor rank. In particular, women in science and social science are less likely to have tenure track jobs within five years of the doctorate when compared with men. However, when controls for marital status and children are included in the analysis, the research finds that unmarried women are significantly more likely to have tenure track jobs than unmarried men. Marriage provides a significant advantage for men relative to women. Presence of children, especially young children, significantly disadvantages women while having no impact on men in obtaining tenure track jobs. The research also finds no significant gender differences in the probability of obtaining tenure in life science, physical science, and engineering. These results also hold for promotion to full professor. However, significant gender promotion differences are evident in the social sciences, in particular, economics. Finally, the research finds large gender differences in salaries are partially explained by academic rank. However, gender salary differences for full professors, on the order of 13% in the sciences, are not fully explained by observable characteristics.

In his examination of the salaries and appointments of men and women in academia, the Director of Research at the American Association of University Professors (AAUP) observes: "Substantial disparities in salary, rank, and tenure

---

*Paper presented at the National Academies Convocation on Maximizing the Success of Women in Science and Engineering: Biological, Social, and Organizational Components of Success, held December 9, 2005, in Washington, DC. I thank the National Science Foundation for granting a site license to use the data and Kelly Kang of the NSF for providing technical documentation. Ronnie Mukherjee provided research assistance. The use of NSF data does not imply NSF endorsement of the research, research methods, or conclusions contained in this report. Financial support was provided from NSF grant SES-0353703. Any errors are my own responsibility.

between male and female faculty persist despite the increasing proportion of women in the academic profession" (Benjamin, 1999). While the evidence presented by AAUP is striking, the gender comparisons of salaries do not control for characteristics that contribute to pay differentials such as academic field or publication record. Simply comparing salaries of male and female academic scientists without taking into consideration these factors could overstate the gender salary gap. Disentangling the causes of gender disparities in employment outcomes requires an in-depth examination of the data. This report summarizes research that examines the relationship between hiring, promotion, and salary for tenure track faculty using data from the Survey of Doctorate Recipients (SDR).

The Economic Perspective

Economic theory provides the underpinnings of this research. I start by assuming that employment outcomes are determined by market forces. Wages and hiring are determined by the supply of and demand for PhD scientists. Equally productive workers irregardless of gender will be paid the same and hired in similar numbers given market forces. Given these assumptions, one should not observe hiring, promotion, and salary differences for equally productive workers of either gender. However, persistent gender wage and employment differentials persist on average in the market as a whole (Altonji and Blank, 1999) and for scientists in particular (Ginther, 2001). I use economic theory to explain observed gender differences in hiring, promotion and salary.

Beginning with Becker's seminal work on discrimination (Becker, 1971), economists have developed models to understand gender and racial disparities in employment outcomes. Becker argues that taste-based discrimination (prejudice) will be eliminated by competitive forces. Given employer, employee, or customer prejudice, those firms that pay premiums to favored workers will have higher costs. Thus, the nondiscriminating firm will have a competitive advantage by hiring women or minorities, and the market will eventually compete away the discriminating wage differential. Becker's prediction relies on the assumption that markets are perfectly competitive—an assumption one can reject for academic institutions.

Given Becker's results, economic theory has developed other explanations besides discrimination to account for observed gender differences in employment outcomes. These explanations may be divided into differences in "preferences" or choices and other factors. The preference-based explanations argue that gender differences in employment outcomes result from choices, in particular, differences in productivity. Economic theory holds that equally productive workers will be paid the same, thus, gender salary differences are the result of differences in productivity. A second preference-based explanation is that women chose to marry and have children, which in turn affects their attachment to their careers and overall productivity.

Other theoretical explanations include monopsony models of the labor market. A monopsonist is a single employer of labor that has more bargaining power in the employment contract than the worker. Monopsonists pay workers less than the competitive wage and may be able to pay different wages to different types of workers depending upon their relative mobility. Thus if female faculty have fewer outside job opportunities, this will generate a gender wage differential. One may convincingly argue that academic institutions have monopsony power relative to faculty in most fields. However, for monopsony to explain gender employment disparities, women would need to be less mobile than men.

Job-matching models may also explain gender differences in employment outcomes. In this model workers who are the best match for the job earn the highest salaries. In loose terms, the job-matching model suggests that women are paid less because they are not as capable (not as good of a match) in science compared to men.

If the researcher cannot explain the gender differences in employment outcomes using one of the above explanations, then the residual gender difference in hiring, promotion, or salary may be attributed to discrimination. Statistical discrimination suggests that imperfect information on the part of employers generates wage differentials. In this model, an employer attributes the average characteristics of a group to an individual member of this group—essentially the employer uses a stereotype in making hiring decisions or setting wages. As a result, we observe gender differences in employment outcomes. However, direct measures of statistical discrimination are difficult to come by. Thus, discrimination may be inferred when other plausible explanations have been ruled out.

Using economic theory as a guide, the research summarized in this report is organized using three basic principles. First, there is no single scientific labor market. As a result, this research disaggregates the data by scientific field. Second, gender differences in employment outcomes need a context in order to make meaningful comparisons. Thus, the research compares employment outcomes across academic fields in order to ascertain the relative status of women in academic science and social science. Finally, employment outcomes are interrelated. One cannot understand gender differences in salary without considering related outcomes of hiring and promotion. Given these principles, my research poses the question: Does science discriminate against women? I evaluate gender differences in hiring, promotion, and salary and can largely explain the first two outcomes using observable characteristics. However, I find large gender differences in the salaries of full professors that I cannot explain as a function of productivity or other choices.

Data and Methods

This study uses data from the Survey of Earned Doctorates (SED) and the Survey of Doctorate Recipients (SDR) to examine the distribution of women across scientific fields and gender differences in salary. The SED is a census of doctorates awarded in the United States each year. I use the 1974–2004 waves of the survey to evaluate changes in the distribution of women in scientific fields. The SDR is a nationally representative sample of PhD scientists in the United States used by the National Science Foundation to monitor the scientific workforce and fulfill its congressional mandate to monitor the status of women in science. This study uses data from the 1973-2001 waves of the SDR. The SDR collects detailed information on doctorate recipients including demographic characteristics, educational background, employer characteristics, academic rank, government support, primary work activity, productivity, and salary. Although the SDR has comprehensive measures of factors that influence academic salaries, the data lack information on some quantitative measures, such as laboratory space and extensive measures of publications. Measures of academic productivity are largely missing from the SDR data, but the SDR does ask questions about publications in the 1983, 1995, and 2001 surveys. I use these data to create rough measures of productivity for each year following the doctorate.[1]

Academics in the life sciences, physical sciences, engineering, and social science are included in the analysis. Life science includes biological sciences and agriculture and food science. Physical science includes mathematics and computer science, chemistry, earth science and physics. Social science includes economics, psychology, sociology and anthropology, and political science. Engineering includes all engineering fields. The SDR collected information on doctorate recipients in the humanities between 1977 and 1995. In some of the analysis that follows, I include comparisons across the three broad disciplines of humanities, sciences, and social sciences.

I begin the analysis by analyzing the percentage of doctorates awarded and the percentage of tenured faculty who are female. Figures 2-1 and 2-2 indicate that women are not equally distributed across scientific fields. Figure 2-1 graphs the percentage of doctorates awarded to females between 1974 and 2004 using data from the SED. If we consider only life science fields, we may conclude, like the National Research Council (2001), that women have indeed moved 'from scarcity to visibility' in terms of doctorates granted. By 2004 almost half of all doctorates in life science and more than half of all doctorates in social science were awarded to women. However, both physical science and engineering awarded less than one-third of doctorates to women. In the year 2004, less than 18% of engineering doctorates and less than 27% of physical science doctorates were granted to women.

Despite the increasing numbers of doctorates awarded to women, the representation of women among tenured academic scientists remains quite low. Figure 2-2 uses data from the 1973–2001 waves of the SDR to graph the percentage of

---

[1]Specifics of the data creation may be found in Ginther (2001) and Ginther and Kahn (2005).

SECTION 2: SELECTED WORKSHOP PAPERS   103

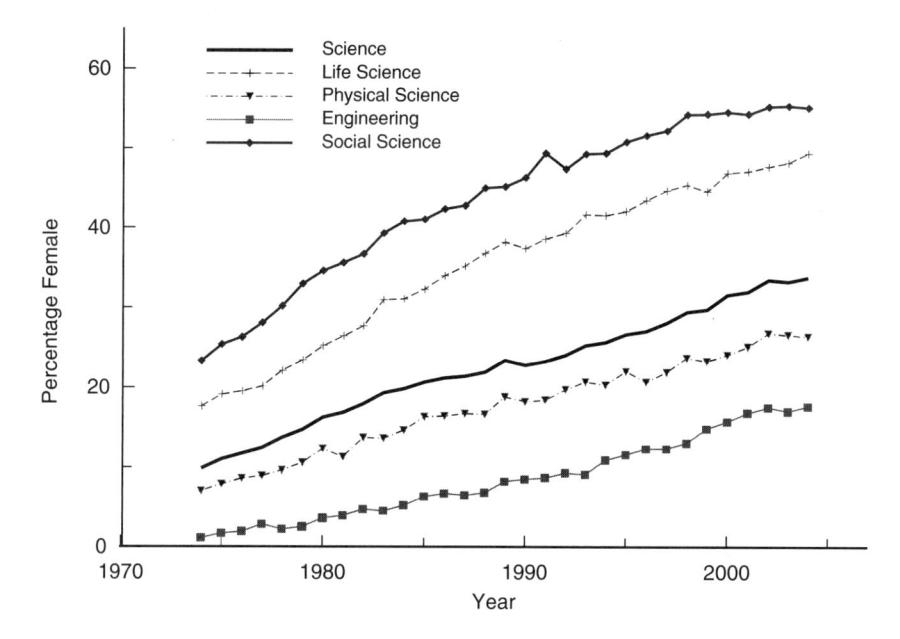

FIGURE 2-1 Percentage of doctorates granted to females, 1974–2004.
SOURCE: 1974-2004 Survey of Earned Doctorates.

tenured faculty who are female in life science, physical science, social science, and engineering. As expected, social science and life science have the highest percentages of tenured female faculty at 28 and 25% respectively in 2001. Physical science and engineering have far fewer tenured female faculty at 11 and 5%, respectively. Given the large differences between the percentages of doctorates awarded to women and the percentages of tenured faculty who are women, I turn to potential explanations.

Gender Differences in Hiring and Promotion

Hiring

The underrepresentation of women in tenured academic ranks may result from gender differences in hiring or promotion. Ginther and Kahn (2005) examine gender differences in hiring by evaluating whether women in science are more or less likely than men to get tenure track jobs within five years of receiving their doctorate. Women and men who leave academia immediately following the doctorate are dropped from the sample. Figure 2-3 shows three sets of estimates of the effect of being female on getting a tenure track job using samples of over

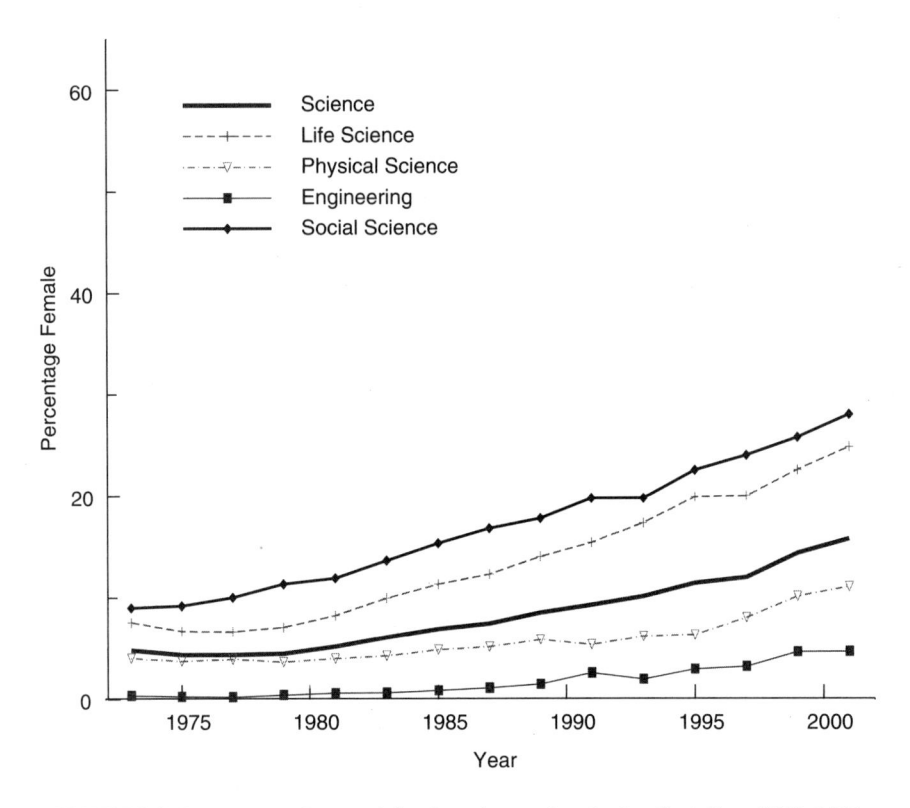

FIGURE 2-2 Percentage of tenured faculty who are female, by discipline, 1973–2001.
SOURCE: 1973-2001 Survey of Doctorate Recipients.

12,000 scientists and over 3,000 social scientists from 1973–2001. Negative numbers indicate that women are less likely whereas positive numbers indicate that women are more likely to get a tenure track job within five years of PhD. Numbers that are underlined are statistically significant at the 5% level. The first bar in Figure 2-4 shows that women are between 4 to 6% less likely than men to have tenure-track jobs in all science fields combined, social science, and life science. There is no significant difference between men and women getting a tenure-track job in physical science and engineering. The second bar in Figure 2-4 includes controls for academic field, race, age at PhD, year of PhD, marital status, and children. The estimated gender gap falls for all science and social science fields but does not change appreciably for the disaggregated science fields.

The third bar includes controls that interact female with marital status and children. These interaction terms allow the impact of marriage and children to be different for men and women in the model. The estimates are strikingly different.

SECTION 2: SELECTED WORKSHOP PAPERS 105

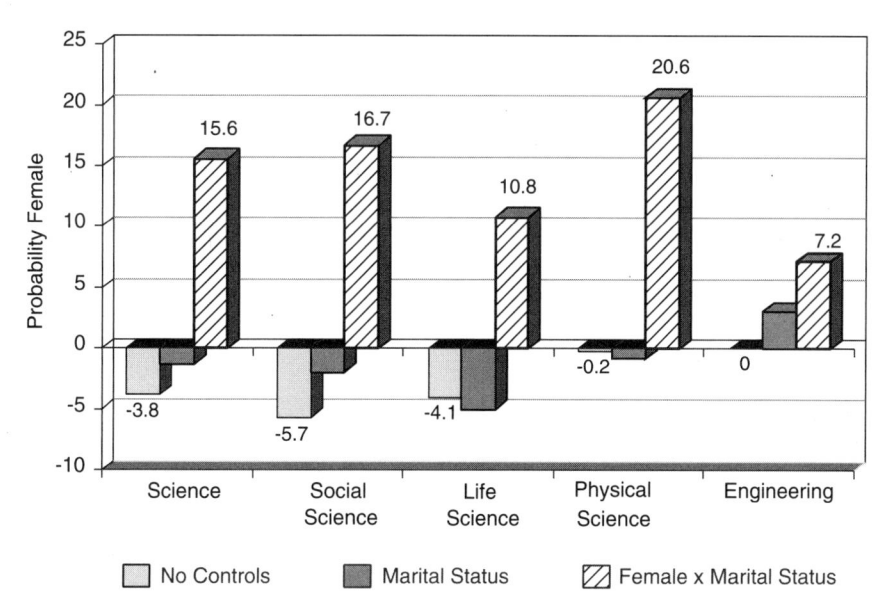

FIGURE 2-3 Gender differences in tenure-track job within 5 years of PhD.
Notes: Estimates from Ginther and Kahn (2005) using 1973-2001 Survey of Doctorate Recipients.

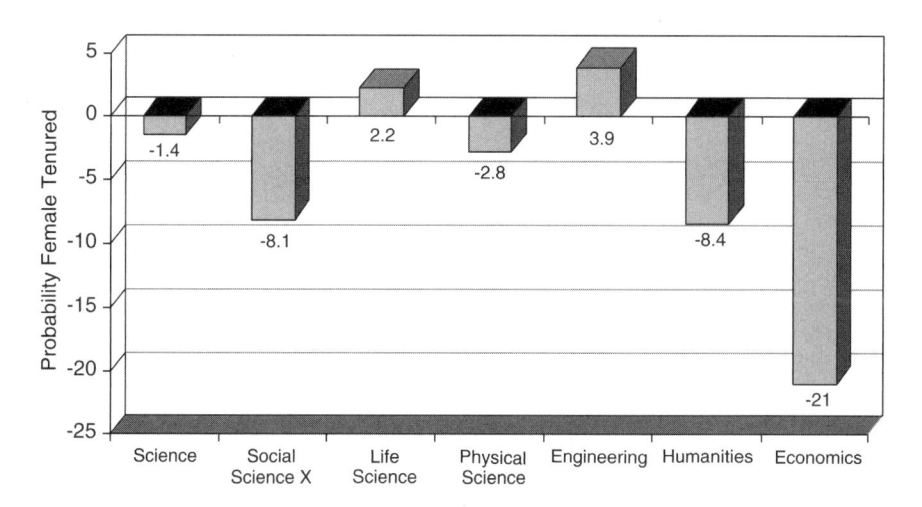

FIGURE 2-4 Gender differences in promotion to tenure 10 years past PhD.
Notes: Estimates from Ginther and Kahn (2004) and Ginther and Hayes (2003). Science and Social Science estimates from 1973-2001 SDR. Humanities estimates from 1977-1995 SDR. Economics, humanities, and social science X (excluding economics) are statistically significant (p = 0.01).

Women are between 7 to 21% *more likely* than men to get a tenure-track job within 5 years of PhD provided they are unmarried and do not have children. These results indicate that much of the underrepresentation of women in academic science is the result of having children. Single women are 16% more likely in science and 17% more likely in social science to get tenure-track jobs than single men. Marriage has a positive and significant impact of 22% on men getting a tenure-track job whereas the effect of marriage on women ranges between 0 and 8% for all science, life science, and social science fields. The exception is engineering where marriage increases women's chances of having a tenure-track job by 23%. Children, especially young children, significantly decrease the likelihood of women obtaining a tenure-track job between 8 to 10% in all science fields, life science, and social science while having no significant impact on men.

The positive impact of marriage and children on men's tenure-track employment echoes the positive impact of men's marriage and children on wages and promotion in the labor market as a whole. The negative impact of children on women's tenure-track employment may result from a number of factors. Women may choose to have children instead of pursuing an academic career because of the coincident timing of the tenure and biological clocks. The dual-career problem may also play a role. Career hierarchies in marriage often result in the husband's career taking precedence over the wife's career. If it is difficult to obtain two tenure-track jobs, she may choose to have children instead of investing in her career.

Furthermore, women are often the primary caregivers of children and this may hamper investments in their careers. The availability of tenure-track jobs may be limited to such an extent that women choose to invest more in marriage and family than in their careers. I suggest that the relative lack of academic jobs may be playing a significant role. By way of example, approximately half of all medical students are women and increasing numbers of women are practicing medicine. The demand for doctors is much higher than the demand for academic scientists, and this demand results in more women practicing medicine. It follows that the lack of academic jobs may be contributing to women's underrepresentation in academic science.

Finally, the timing of women's departure from academia may also indicate problems with the post-doctoral system in academic science. Studies suggest that the post-doctoral process is taking longer because the number of post-doctoral positions has expanded without a similar expansion of academic jobs (Davis, 2005). These results suggest that some combination of factors at the early stages of women's careers are affecting married women's choice of or access to tenure-track jobs. I now examine what happens to women as they progress through the tenure track.

Promotion

Once women have tenure-track jobs, their prospects for getting tenure in science are very promising but less so in social science. Figure 2-4 is derived from estimates in Ginther and Kahn (2004, 2005) and Ginther and Hayes (2003). It shows gender differences in the promotion to tenure 10 years past the doctorate in the fields of science, social science excluding economics, life science, physical science, engineering, humanities, and economics. These latter two disciplines are included to provide a context for the remaining fields. Women are between 1 to 3% less likely to get tenure in all scientific fields combined and in physical science 10 years past the doctorate. Women are between 2 and 4% *more likely* to get tenure in life science and engineering. These results indicate that gender differences in promotion to tenure are small for women in scientific fields.

This is not true for social science (excluding economics) and the humanities where women are 8% less likely than men to get tenure. Economics is the outlier—women are 21% less likely to get tenure than men 10 years past the doctorate. These differences in economics cannot be fully explained by gender differences in productivity, marital status, or presence of children (Ginther and Kahn, 2004).

Ginther and Kahn (2005) estimate gender differences in promotion to tenure and promotion to full professor in scientific fields. They find no statistically significant gender differences in promotion to either rank. Thus, we can conclude that gender differences in promotion in science are negligible. However, gender differences in promotion in social science are large, especially in economics. I now consider gender differences in salaries.

Gender Differences in Salaries

There are several factors that affect the salaries of academics. Demographic characteristics such as race, marital status, fertility, and years of work experience may have a positive or negative effect on salaries. For example, on average, marriage increases male salaries while having a negative effect on female salaries. Employer characteristics such as working at a public or private institution, liberal arts or a doctoral institution, and the Carnegie ranking of the employer may also affect salaries. Top research institutions pay more than liberal arts colleges. Public institutions have state-mandated salary scales that tend to be more restrictive than those at private institutions. Employee characteristics such as the academic rank and tenure status of the individual also influence salaries, with salaries increasing with academic rank and tenure.

Measures of productivity also affect salaries. These include factors such as whether the individual receives government support, primary work activities, and publications. If men are more likely to work at top-ranked research universities, the gender salary gap will be larger. Salary differences may also result from differential treatment reflected in differences in estimated coefficients. For example, at private institutions if men are paid more than women and private institutions are equally likely to employ both, then the gender salary gap will increase. Taken together, these observable characteristics may explain a substantial portion of the gender salary gap.

The analysis reported here updates estimates in Ginther (2001, 2003, 2004) and Ginther and Hayes (2003) using the 2001 SDR data. The first bar in Figure 2-5 shows the average gender salary gap for all tenure-track and tenured faculty combined in science, social science, life science, physical science, engineering, and humanities. The salary gap ranges from a low of 11% in the humanities[2] to a high of 21% in engineering. This combined gender salary gap is very large. However, previous research by Ginther and Hayes (1999, 2003) shows that the majority of the gender salary gap in the humanities disappears when separate salary regressions are estimated for each academic rank.

The remaining bars in Figure 2-5 show the gender salary gap for assistant, associate, and full professor ranks. Similar to Ginther and Hayes (1999, 2003), the gender salary gap at the assistant and associate professor ranks falls from close to 20% to just over 5% for assistant and associate professors in science and social science. However, the full professor salary gap increases to 8% for social science and as high as 14% for life scientists. In contrast, the gender salary gap for full professors in the humanities is less than 2%.

Using regression techniques, these salary gaps can be decomposed into factors that are explained by observable characteristics and factors that result from differential treatment of men and women. One-third of the salary gap for all science fields combined cannot be explained by observable characteristics such as productivity. Three-quarters of the salary gap for engineering cannot be explained by observable characteristics. I now evaluate whether economic theory can explain the gender salary gap for full professors.

Explanations for the Salary Gap

To determine whether publication differences could account for a substantial portion of the unexplained salary gap for full professors, I use publications measures from the 2001 SDR (Ginther, 2004). The sample includes measures of papers published and papers presented at conferences within the last five years. Including productivity measures only reduced the unexplained portion of the gap by 0.3 percentage points from 3.8 to 3.5%. Thus, productivity does not appreciably reduce the unexplained gender salary gap for full professors for all science fields combined. However, productivity differences do explain a significant portion of the salary gap in physical science and engineering.

---

[2]This estimate is based on 1995 SDR data, the last year information on the humanities was available.

SECTION 2: SELECTED WORKSHOP PAPERS 109

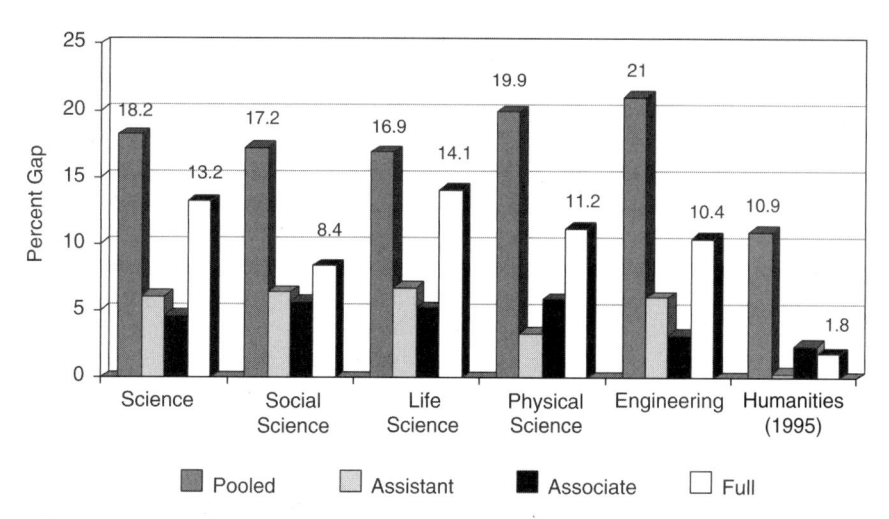

FIGURE 2-5 Gender salary gap by academic rank, 2001 SDR.
Notes: Estimates for Humanities from Ginther and Hayes (2003) based on 1995 SDR.

Next, I consider other factors that may explain the gender salary gap. In particular, women who have children are often paid less than women without children (Waldfogel, 1998). Since women are often the primary care-givers for children, having a child may reduce a woman's productivity. My analysis shows that the total number of children and presence of children under the age of six have little or no impact on either the explained or unexplained portion of the gender salary gap for full professors.

Economic models of monopsony (where the university acts as the sole purchaser of labor) may also explain the gender salary gap. In monopsonistic models of academic labor markets developed by Ransom (1993), senior faculty have higher moving costs and receive lower salary offers. It is possible that tenured women faculty have higher moving costs than their male colleagues because of dual career considerations or fewer job opportunities. In related research, Booth, Frank, and Blackaby (2002) suggest that universities may consider women to be "loyal servants" who are less likely to change academic employers. As a result, universities can make lower salary offers and adjustments to women scientists. Both the monopsony and loyal servant explanations would be evident in the effect of job tenure on wages. If women have higher moving costs due to monopsony or are perceived to be "loyal servants," their wages would be reduced more than men's for each additional year of job tenure with the same employer. However, the data show the opposite is true. Male salaries are reduced more than female salaries for each additional year of job tenure. Thus, neither monopsony models nor the loyal servant hypothesis provide an adequate explanation of the gender salary gap in science.

Job matching models suggest that women are paid less than men because they are not as well suited (matched) to scientific careers. Whereas this may explain part of the salary gap for lower ranks, it is difficult to argue that women full professors of science are not well suited to academic science.

Although productivity, children, and economic models do not provide an adequate explanation for the gender salary gap, there are other variables that are associated with the gender gap. In my analysis, the single most important factor contributing to both the explained and unexplained gender gap is work experience—measured by years since PhD. Virtually all of the explained salary gap for full professors results from men having relatively more work experience. In addition, virtually all of the unexplained salary gap for full professors results from men having a higher return on experience than women. Although the effect of experience on wages is almost the same for men and women in the assistant and associate professor ranks, it differs for men and women at the full professor rank. Each additional year of work experience increases the salaries for male full professors but has zero effect on the salaries of female full professors, thus contributing to the unexplained salary gap.

The effect of experience suggests that the gender salary gap may result from a subtle mechanism such as the cumulative advantage model described by Zuckerman (1987). In this model, some groups receive greater opportunities than others. Recipients are enriched and nonrecipients are impoverished. Over time as advantages and disadvantages accumulate, a gender gap develops. The estimated impact of experience on the salary gap is consistent with the cumulative advantage model.

Conclusions and Policy Recommendations

I began this analysis by posing the question: does science discriminate against women in hiring, promotion, and salaries? The answers to these questions provide questions for further research and policy recommendations.

First, women in science and social science are less likely to have tenure track jobs within 5 years of the doctorate when compared with men. However, when controls for marital status and children are included in the analysis, the research finds that unmarried women are significantly *more* likely to have tenure track jobs than unmarried men. Marriage provides a significant advantage for men relative to women. Presence of children, especially young children, significantly disadvantages women while having no impact on men in obtaining tenure track jobs. Second, the research finds no significant gender differences in the probability of obtaining tenure in life science, physical science, and engineering. These results also hold for promotion to full professor. However, significant gender promotion differences are evident in the social sciences, in particular, economics. Finally, the research finds large gender differences in salaries are partially explained by academic rank. However, gender salary differences for full professors, on the order of 13% in the sciences, are not fully explained by observable characteristics. The gender differences in salaries are most consistent with the cumulative advantage model where advantages accrue to men more often than women and generate salary differentials.

The results of this research provide both research and policy recommendations. The gender differences in hiring and salary summarized in this paper can only be partially explained with existing data. In order to understand the complex causes of gender disparities in employment outcomes for women in science and social science, better data are required. The Survey of Doctorate Recipients is the best source of data on academic labor markets. However the quality of the data should be enhanced along two dimensions. First, additional questions should be included in the SDR to allow for the comparison of resource allocations. These questions include the following:

- Information on publications and citations
- Dollar amount and duration of grant awards
- Laboratory size
- Numbers of graduate students and post-doctoral students advised.

This series of questions would allow researchers to determine whether gender differences in resource allocation and productivity contribute to the gender salary gap.

Second, additional questions related to post-doctoral appointments and dual career issues should be include in the SDR. These questions include:

- Number, quality, and productivity of post-doctoral appointments
- Spouse information including education, employment and earnings
- Childcare time

This series of questions would allow researchers to determine whether the post-doctoral process or work-family trade-offs lead to fewer women in academic science.

In addition to the SDR, I recommend that agencies such as the NSF and NIH collect information on the demand for scientists. In particular, researchers could make great use of data on the number of academic and nonacademic jobs available in scientific fields. It is my belief that the excess supply of scientists in certain fields disproportionately disadvantages women. Finally, I recommend that the NSF create an advisory panel of researchers who use the SDR to make recommendations on data collection, survey design, survey questions, and dissemination of the data.

The hiring and salary gaps summarized in this research also lead to specific policy recommendations. In terms of hiring, universities should be encouraged to develop family friendly policies such as tenure clock stops for childbirth, paid parental leave, and on-site childcare. These policies would ease the burden of having and caring for children. Dual career hiring policies may also benefit women. At most institutions, accommodations for the trailing spouse are ad hoc or nonexistent. This poses a special problem for women who are more likely to married to professional or academic spouses. Universities that wisely invest in academic couples may be able to hire and retain higher quality faculty because couples are less mobile than individuals. Finally, I would recommend institutional review of salaries on a regular basis in order to adjust obvious gender salary discrepancies.

References

JG Altonji and RM Blank (1999). Race and Gender in the Labor Market. *Handbook of Labor Economics, Volume 3*, Eds. O Ashenfelter and D Card. Amsterdam: Elsevier Science.

GS Becker (1971). *The Economics of Discrimination*, 2nd edition. Chicago: University of Chicago Press.

E Benjamin (1999). Disparities in the salaries and appointments of academic women and men. *Academe* 85(1):60-62.

AL Booth, J Frank, and D Blackaby (2002). "Outside Offers and the Gender Pay Gap: Empirical Evidence from the UK Academic Labour Market". Mimeo, University of Essex.

G Davis (2005). "The Productive Postdoc: Assessing the Impact of Recommended Changes to the Postdoctoral Experience". Mimeo, Sigma Xi.

DK Ginther (2001). Does Science Discriminate Against Women? Evidence from Academia, 1973-97. *Federal Reserve Bank of Atlanta Working Papers 2001*-02 (2001):66. http://www.frbatlanta.org/publica/work_papers/wp01/wp0102.htm.

DK Ginther (2003). Is MIT the exception? Gender pay differentials in academic science. *Bulletin of Science, Technology, and Society* 23(1):21-26.

DK Ginther (2004). Why women earn less: Economic explanations for the gender salary gap in science. *AWIS Magazine* 33(1):6-10.

DK Ginther and KJ Hayes (1999). Gender differences in salary and promotion in the humanities. *American Economic Review Papers and Proceedings* 89(2):397-402.

DK Ginther and KJ Hayes (2003). Gender differences in salary and promotion for faculty in the humanities, 1977-1995. *The Journal of Human Resources* 38(1):34-73.

DK Ginther and S Kahn (2004). Women in economics: Moving up or falling off the academic career ladder? *Journal of Economic Perspectives* 18(3):193-214.

DK Ginther and S Kahn (2005). Does Science Promote Women? Evidence from Academia 1973-2001. Mimeo, University of Kansas.

National Research Council (2001). *From Scarcity to Visibility*. Washington, DC: National Academy Press.

M Ransom (1993). Seniority and monopsony in the academic labor market. *American Economic Review* 83(1):221-233.

SV Rosser (2004). *The Science Glass Ceiling*. New York: Routledge.

J Waldfogel (1998). The family gap for young women in the United States and Britain: Can maternity leave make a difference? *Journal of Labor Economics* 16(3):505-545.

H Zuckerman (1987). "The Careers of Men and Women Scientists: A Review of Current Research." Reprinted in eds. H Zuckerman, JR Cole, and JT Bruer, *The Outer Circle: Women in the Scientific Community*. New York: WW Norton.

BIOPSYCHOSOCIAL CONTRIBUTIONS TO COGNITIVE PERFORMANCE*

Diane F. Halpern
Berger Institute for Work, Family, and Children
Claremont McKenna College

Abstract

Females and males are both similar and different in their cognitive performance. There is no evidence to support claims for a smarter sex. Males and females have different average scores on different cognitive measures; some show an advantage for females and others show an advantage for males. Females are achieving at higher rates in school at all levels and in all subjects, including subjects in which they obtain lower scores on aptitude/ability tests (e.g., advanced mathematics). Although there is much overlap in the female and male distributions, on average, females excel on many memory tasks including memory for objects and location, episodic memory, reading literacy, speech fluency, and writing. Males excel at visuospatial transformations, especially mental rotation, science achievement, mathematics tests that are not tied to a specified curriculum (possibly due to use of novel visuospatial representations and transformations), and males are more variable on many cognitive tests. A biopsychosocial model that recognizes the reciprocal relationships among many types of variables is used as an explanatory framework.

There have been remarkable changes in the lives of women and men in the blink of history that was the 20th century. College enrollments went from consisting largely of men from the privileged classes near the start of the 20th century to men from all socioeconomic classes and literally, all stripes, as they returned from World War II near mid-century. College enrollments for women at the same time consisted mostly of women of privilege, or exceptional talent, or high moti-

---

*Paper presented at the National Academies Convocation on Maximizing the Success of Women in Science and Engineering: Biological, Social, and Organizational Components of Success, held December 9, 2005 in Washington, DC.

Some authors prefer to use the term "gender" when referring to female and male differences that are social in origin and "sex" when referring to differences that are biological in origin. In keeping with the biopsychsocial model that is advocated in this paper and the belief that these two types of influences are interdependent and cannot be separated, only one term is used in this chapter. "Sex" is used without reference to the origin of any observed differences or similarities and is not meant to imply a preference for biological explanations. These terms are often used inconsistently in the literature.

vation, or some combination of all three. But, by the time the post-war baby boom reached college age, women were attending college at an increasingly higher rate than earlier generations, in part because the baby boomers faced more competition as they entered an overcrowded work force. By 1982, the number of women enrolled in and graduating from college exceeded that of men, and the gap in favor of women has continued to widen ever since.

Among women between 25 and 34 years old, 33% have completed college, compared to 29% of men. Women also get higher grades in school, on average, in every subject area (Dwyer and Johnson, 1997; Kimball, 1989). These changes have occurred faster than any gene can mutate or any theory of evolution can explain, so it is not surprising that most people look to societal explanations for the changing roles of men and women. Although women still dominate enrollments in the "helping professions," such as teaching, social work, and nursing, they have been increasing their enrollments in traditional male disciplines. Males have been much slower to enter the traditional female disciplines. There have been many initiatives to accelerate the increase in the numbers of women in academic areas commonly known as STEM—Science, Technology, Engineering, and Mathematics—however the underrepresentation of women, particularly at the full professor level in university faculties, was brought into a near frenzy of public debate when Lawrence Summers (January 14, 2005), president of Harvard University, offered his personal beliefs about this topic. Summers identified these three broad hypotheses as possible reasons for the large disparities in the percentage of women in academic positions in universities: (1) high-powered job hypothesis; (2) differential availability of aptitude at the high end, and (3) different socialization and patterns of discrimination in the faculty search process. Summers eliminated the third hypothesis quite simply by concluding that there could not be discrimination against women in the process of searching and hiring professors because discrimination would have to occur on every campus in the United States. If there were one or even a few campuses that did not discriminate against women scientists, then these campuses would have many outstanding women at the level of full professor who had been discriminated against at the other campuses; since there are no such campuses, there could not have been discrimination in the hiring or promotion process. Summers' hasty dismissal of all that is known about implicit stereotyping (Banaji and Hardin, 1996), social expectations, in-group and out-group behaviors (Shelton and Richeson, 2005), and social psychology created a firestorm of controversy. He later retracted his statements and pledged $50 million to enhance faculty diversity and support women's programs at Harvard. The other two hypotheses proposed by Summers are addressed in greater detail below.

Summers' statements raised a serious question that is often not asked at the many symposia and talk shows that have followed as a result of his remarks: Are there too few women with the cognitive abilities to become our highest level scientists and mathematicians?

There are many science disciplines and women are dominating some of them. Women now comprise 75% of all graduating veterinarians, a field that is sometimes considered one of the most difficult of the sciences because there are multiple biological systems to be learned; women are obtaining 50% of medical school degrees, and 44% of the PhDs in the biological and life sciences, so women clearly have the innate ability to succeed in science. By contrast, women are getting only 29% of the doctorates in mathematics; 17% in engineering; and 22% in computer/information sciences. These percentages are higher than they used to be, but not equal to the number of males in these areas. On the other hand, should we be just as concerned about the low percentage of men who obtain only 32% of PhDs in psychology, 37% in health sciences, 34% in education (U.S. Department of Education, 2000)? Clearly women have the cognitive ability to learn and succeed in math and science, although there are sex differences in the fields of sciences in which they are selecting. The differences among these fields are sometimes described by a theory that suggests that biological or life sciences are preferred by women and inorganic sciences are preferred by men, but when psychologists look over this list, alternative categorizations emerge. For example, Lippa (1998) found that women, by a large margin, prefer to work with people—a career preference that also fits with women's success in the field of law, which used to be dominated by men, versus men's, strong preference for working with "things." Ackerman et al. (2001) studied how trait complexes, which consist of abilities, interests, and personality variables, combine to influence achievement and career goals.

These data raise interesting philosophical questions about values and opportunities: would we expect or want all fields of study and all careers to become approximately equal in the numbers of men and women, and if so, at what cost are we willing to pursue that goal?

Biopsychosocial Model

When it comes to understanding cognitive performance, males and females are both similar and different, and some of the differences are small and some are large. There are cognitive tasks and tests that show, on average, some differences that favor females and some that favor males. There is also much overlap, so we do not have distinctly different groups, but overlapping distributions. In thinking about the differences, some of them have not changed over the decades for which we have data. Most people prefer environmental explanations, but are willing to settle for an explanation the will give a percentage of the "explanation" to nurture, a percentage to nature, and a percentage to their interaction. But nature and nurture cannot act independently, and they cannot "just interact." Nature and nurture mutually influence each other in reciprocal ways and cannot be separated. It is not as though there is a number that exists in the real world and if researchers are very clever they will discover the percentage that can be attributed to nature or nurture and their interaction. Nature and nurture have no meaning without each other—nature needs nurture and vice versa.

The distinction between biology and experience is hopelessly blurred, so asking whether nature or nurture plays the greater part in determining a cognitive sex difference is the wrong question. Consider, for example, the brain. It is the quintessential "biological" organ, yet, it is also shaped extensively by experience. There are many sex differences in the architecture of the brain, but it cannot be assumed that differences in female and male brains result solely from genetic or hormonal action. The importance of experience was demonstrated in a study of London cab drivers that found that the cabbies had enlarged portions of their right posterior hippocampus relative to a control group of adults whose employment required less use of spatial navigational skills (Maguire et al., 2000). The cab drivers showed a positive correlation between the size of the region of the hippocampus that is activated during recall of complex routes and the number of years they worked in this occupation. The finding that size of the hippocampus varied as a function of the number of years spent driving taxis makes it likely that it was a lifetime of complex way-finding that caused the brain structure used in certain visual-spatial tasks to increase in size.

The burgeoning field of hormone replacement therapies for men and women is providing evidence that hormones continue to be important in cognition throughout the life span, although the field is complex and rife with controversies. The best evidence for a beneficial effect is the effect of estrogen on verbal memory in old age. Even though there are many studies that have failed to find beneficial effects for hormone replacement in elderly women, a substantial number of studies suggest that exogenous estrogen (pill, patch, cream, or other form) causes positive effects on the cognition of healthy older women and possibly for women in early stages of Alzheimer's disease. This conclusion is in accord with Sherwin's (1999) meta-analytic review of 16 prospective, placebo-controlled studies in humans, where she concludes that "Estrogen specifically maintains verbal memory in women and may prevent or forestall the deterioration in short- and long-term memory that occurs with normal aging. There is also evidence that estrogen decreases the incidence of Alzheimer disease or retards its onset or both" (p. 315). The results of these studies and others provide a causal link between levels of adult hormones and sex-typical patterns of cognitive performance.

A graphic depiction of the biopsychosocial model is shown in Figure 2-6 as a continuous, dynamic loop, essentially blurring the distinction between biology and environment. Learning, for example, is both a biological and environmental variable, with the brain differentially responsive to new learning based on prior learning, genetic factors, nutrition, and much more. Even hormones, which are usually considered "biological" variables, do not act in fixed or preprogrammed ways, but act within a context. We now know, for example, that testosterone can increase or decrease depending on whether an individual wins or loses a game (Schultheiss et al. 2005) and that some cognitive measures vary slightly over the

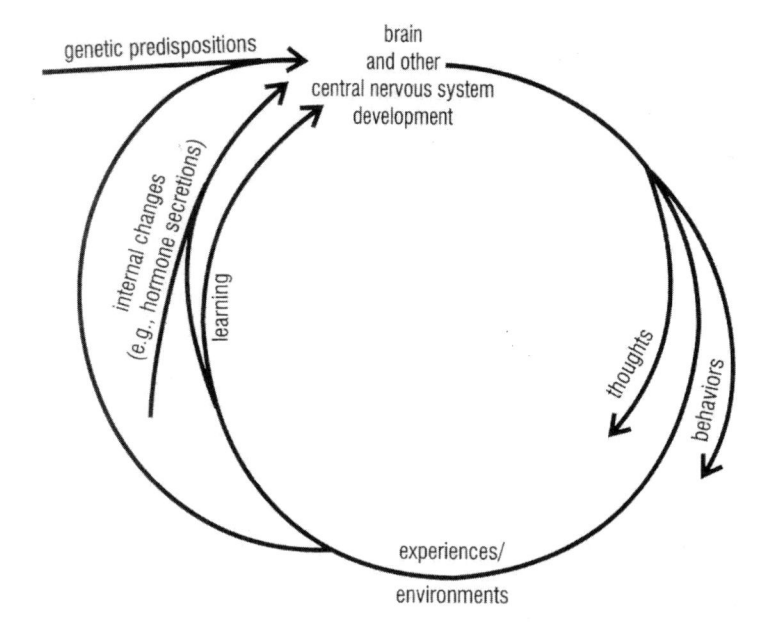

FIGURE 2-6 Biopsychosocial model in which the nature-nurture dichotomy is replaced with a continuous feedback loop.

menstrual cycle for cycling women and over the diurnal cycle for men, but the size of the fluctuations in cognitive performance are too small to be meaningful in everyday life (Halpern and Tan, 2001; Moffat and Hampson, 1996). The biopsychosocial model also makes it easier to understand that although sex differences are often (not always) found on some cognitive tasks, these differences are not immutable or inevitable and "biological" variables are developed in environments that are more or less favorable to their development and maintenance.

Sex Differences in Cognitive Performance

In understanding sex differences in cognitive performance, Hyde's (2005) recent meta-analyses remind us that the sexes are similar in more ways than they are different. The standardized intelligence tests were written and normed to show no overall sex differences, but even a comparison of cognitive tests that were not deliberately normed to eliminate sex differences provide no evidence of overall sex differences in intelligence (Jensen, 1998). These tests do, however, show predictable sex differences on their subscores.

Some researchers object to the study of sex differences because they fear that it promotes false stereotypes and prejudice, but, there is nothing inherently sexist in a list of cognitive sex differences; prejudice is not intrinsic in data, but can be seen in the way people misuse data to promote a particular viewpoint or agenda. Prejudice also exists in the absence of data. Research is the only way to separate myth from empirically supported findings. A necessarily very brief overview of the largest differences is presented here. For a more complete review, see Halpern (2000).

Female:

- **Writing and comprehending complex prose.** In a report published by the U. S. Department of Education, entitled, "Trends in Educational Equity of Girls and Women," the data on reading and writing achievement are described this way, "Females have consistently outperformed males in writing achievement at the 4th, 8th, and 11th grade levels between 1988 and 1996. Differences in male and female writing achievement were relatively large. The writing scores of female 8th graders were comparable to those of 11th grade males" (U.S. Department of Education, 2000, p. 18). A meta-analysis by Hedges and Nowell (1995) called the sex difference in writing that favored girls to be so large as to be "alarming". The female advantage in writing may be one reason why females get higher grades in school, on average. Any assessment that relies on writing provides an advantage to females.
- **Rapid access to and use of phonological, semantic, and episodic information in long term memory.** Many laboratory tests show females are better at generating synonyms, recalling information about events, and numerous standard memory tasks such as object location and identity (Herlitz, Nilsson, & Baeckman, 1997, Levy, Astur, & Frick, 2005).
- **Speech articulation and fine motor tasks.** Females are much less likely to stutter and have better fine motor skills (e.g., O'Boyle, Hoff, & Gill, 1995). These results could be interpreted as females are "naturally" better at typing, or small motor repair, or brain surgery.

Male:

- **Visuospatial transformations, especially mental rotation.** This is a well-replicated and large effect that has not declined in over 30 years (between 0.9 to 1.0 standard deviations; Halpern & Collaer, 2005; Masters & Sanders, 1993; Nordvik & Amponsah, 1998). In addition, performance on mental rotations tasks improve with practice and the improved performance transfers to novel mental rotation stimuli, but performance improves equally for women and men (Peters et al. 1995). Numerous replications with training do not find a sex by training interaction. Females do not especially benefit from training. An example of a mental rotation task is shown in Figure 2-7. The task is to determine if the pairs of figures can be rotated to be identical. When this test is administered on a com-

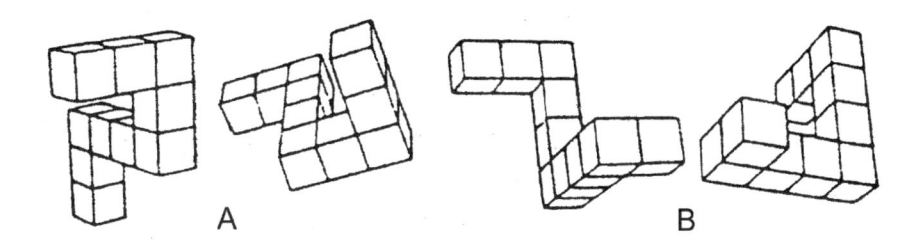

FIGURE 2-7 An example of a mental rotation task. Can the pairs of figures in A and B be rotated so that they are identical? Reaction times and correct answers are recorded.

puter, both the number (and percentage) correct is recorded with the reaction time for each item. Men not only get more items correct, but they also rotate the items more quickly than most women.
- **Fluid (novel) reasoning tasks in math and science.** The advantage for males in mathematics is seen on some math tests. As already noted, females get higher grades in school, even in advanced math and science courses, although there are usually many fewer females enrolled in these courses. The advantage for males in math and science is found on high stakes tests that are not tied to a specific curriculum, which means that the problems require novel approaches, most frequently visuospatial problem representation or transforming visuospatial information in working memory (Gallagher, Levin & Cahalan, 2002). The size of the male advantage gets larger as the population sampled becomes more selective. In other words, the difference between males and females grows larger as the sample moves from high school to college-going students, from college-going students to graduate schools students, and from graduate students to those who are most gifted in math and science among graduate students. As this sample becomes more selective, so does the demand for visuospatial mental representation and transformation, which may be the underlying factor in this cognitive performance differential between males and females.
- **More variable in cognitive performance.** There are more males at both the high and low ends of many cognitive performance distributions. The greater variability for males means that there are more males with mental deficiencies, and there are more males that score at the very high end on many tests of intelligence and achievement. The SAT-M, the mathematics test administered by the Educational Testing Service that is used by many universities for college admissions is one of the tests that shows an excess of males on the extreme high end. The quantitative test of the Graduate Record Examination (GRE-Q), which is used for admissions for graduate school also has a greater proportion of males scoring at its highest end (Webb, Lubinski, & Benbow, 2002).

Distribution of Aptitude

Several researchers have argued that the excess of males at the very high end of the abilities distributions for mathematics can account for the underrepresentation of females in physical sciences and math careers. When Summers referred to the different availability of aptitude at the high end, he was referring the finding that the ratio of males to females in the tails of distributions such as the GRE-Q is very high and gets higher the farther out in the tail that the distribution is cut, so that at the top 1% or 0.5 % there are many more males than females. There are flaws in this line of reasoning as an explanation of the underrepresentation of women in science and math academic careers because there is a lack of females at all ability ranges in science and math, not just at the highest ability range (Halpern, in press). There are many males in science and math who are not in the highest ability ranges because, by definition, only a very small percentage of the population is in this range. In other words, it is not as if we have only mediocre women in sciences and math with a lack at the top—women are underrepresented across the board.

Although the relative scarcity of females in the extreme tails of distributions cannot explain the absence of females in science and math careers overall, a surprising finding showed that for the very highest scoring SAT-M students at age 13, having a "genius" level score made a difference in their own career choices and achievements 20 years later (Wai, Lubinski, and Benbow, 2005). Researchers found that among precocious youth, there were differences in career choices and achievements 20 years later between those youth who scored in the top quartile of the top 1% on the SAT-M and those who scored in the bottom quartile of the top 1% on the SAT-M. Most psychologists would have believed, and probably still believe, that if an individual has achieved a threshold level of ability, additional ability beyond that level has little or no effect on life success because other variables such as motivation, interest, and opportunity would be far more important. These results remind researchers that high level ability is an important determinant of life outcomes, assuming that people have the opportunities to develop their abilities.

In looking over this abbreviated list of areas in which there are cognitive sex differences, one point should be evident—everyone except the profoundly retarded can improve in these cognitive areas with appropriate education, which is why we have schools. We really do not know if we could close, reverse, or increase any or all of the average differences between males and females with learning experiences, "selective breeding" (which was not discussed), hormone manipulations, or with combinations of all of these.

International Comparisons

Some differences between females and males are found consistently in international assessments. International comparisons of males and females are shown in Figure 2-8. The left hand column shows data from 15 year-old students from 25 countries who participated in the Program for International Assessment (PISA). As seen in this figure, all of these countries showed significantly different effects favoring girls in reading literacy. The mathematics achievement and science achievement data are taken from the Third International Math and Science Study (U.S. Department of Education, 1997). The sex differences in math achievement at 8th grade are not as impressive on this assessment as it is on more advanced measures, but as indicated earlier, the size of the sex difference depends on what is assessed and it grows with more select samples. The cross-national consistency of the science achievement data is striking. In looking over these data, it is apparent that the results all show that males performed better than females and that the differences are statistically significant.

Readers may be wondering whether these effects are large enough to be important or meaningful in "real world" contexts. The question of when an effect is large enough to be meaningful has been the subject of much debate. In Valian's (1998) analysis of women's slow advancement in academia and other professions, she showed how small disparities can be compounded over time to create larger disparities, so a seemingly "small" percentage of variance accounted for can be meaningful, depending on the context and variable being assessed. Rosenthal, Rosnow, and Rubin (2000, pp. 15-16), three leading statisticians weigh in on this critical matter: "Mechanically labeling . . . *d*s automatically as 'small,' 'medium,' 'and 'large' can lead to later difficulties. The reason is that even 'small' effects can turn out to be practically important."

In a research paper on the mental rotation test, Peters et al. (1995) report that sex accounted for only 18% of the variance, but when they calculated a Binomial Effect Size Display (BESD), they found that 15% of the females exceed the mean of the males on this test. If the mean value of the male distribution were selected as the cut point for selection for an engineering program or some similar program, 50% of men would be admitted and 15% of women would be admitted, so even a seemingly "small" percentage of variance would have a devastating effect on the number of women admitted to this hypothetical program for further training.

Grades-Tests Disparities

Although females, in general, are doing better in school than their male counterparts (boys are more likely to repeat a grade, be victimized in school, or show up for school unprepared; U.S. Department of Education, 2000), males do better, in general, on standardized tests that are not linked to any specific curriculum, such as the SATs and GREs, which are used for college and graduate school admissions. The grades-tests disparity implies that the SAT-V (verbal) and SAT-M under-predict women's grades in college, which is empirically supported (Cullen, Hardison, and Sackett, 2004). One explanation of the underprediction of women's grades by tests that are not linked to the curriculum is that women are better

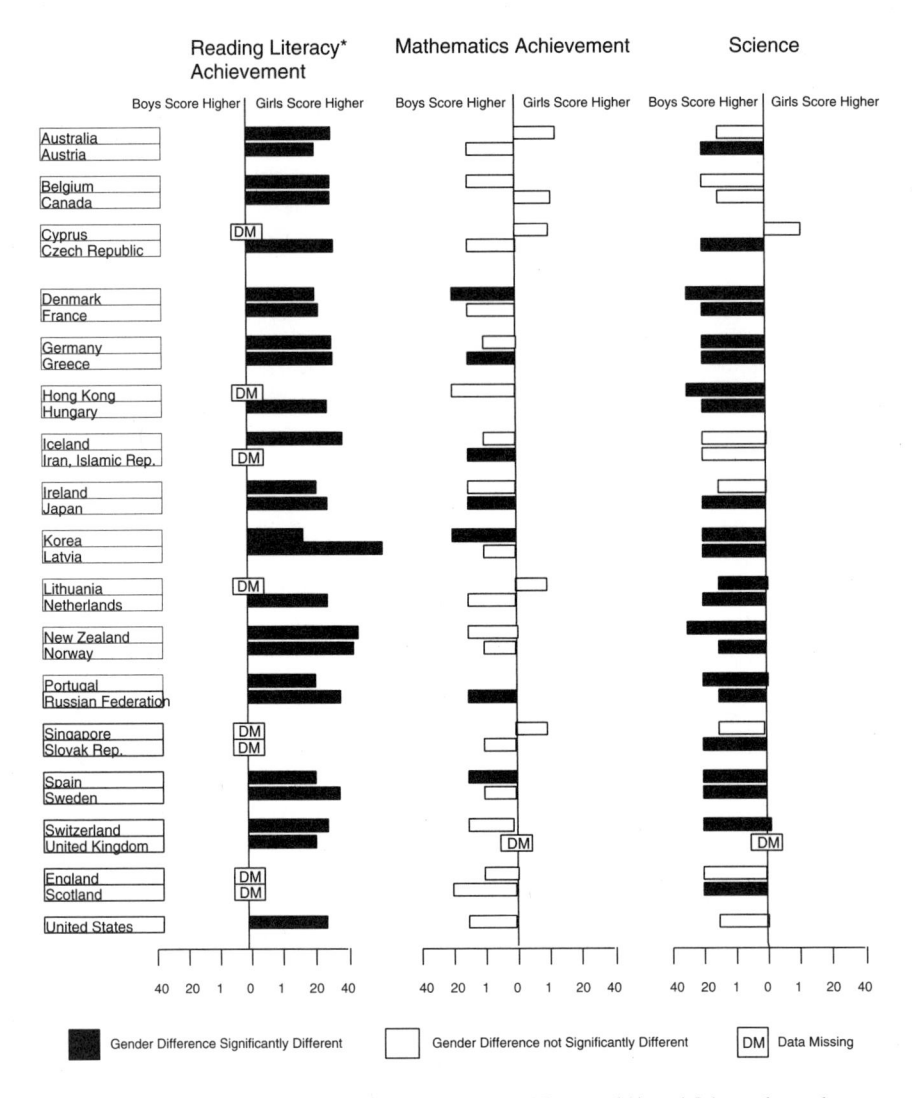

122 COMPONENTS OF SUCCESS FOR WOMEN IN ACADEMIC SCIENCE & ENGINEERING

FIGURE 2-8 Gender differences in achievement: 15 year old* and 8th grade students.

students. Class grades also include classroom behavior and other noncognitive variables that are part of the good student role—a social role that is more compatible with the female sex role than the male sex role.

Average scores on the SAT-M for entering college classes from 1967 to 2004 are shown for men and women are shown in Figure 2-9. Despite the huge changes in number of women enrolled in mathematics courses and their higher grades in

SECTION 2: SELECTED WORKSHOP PAPERS 123

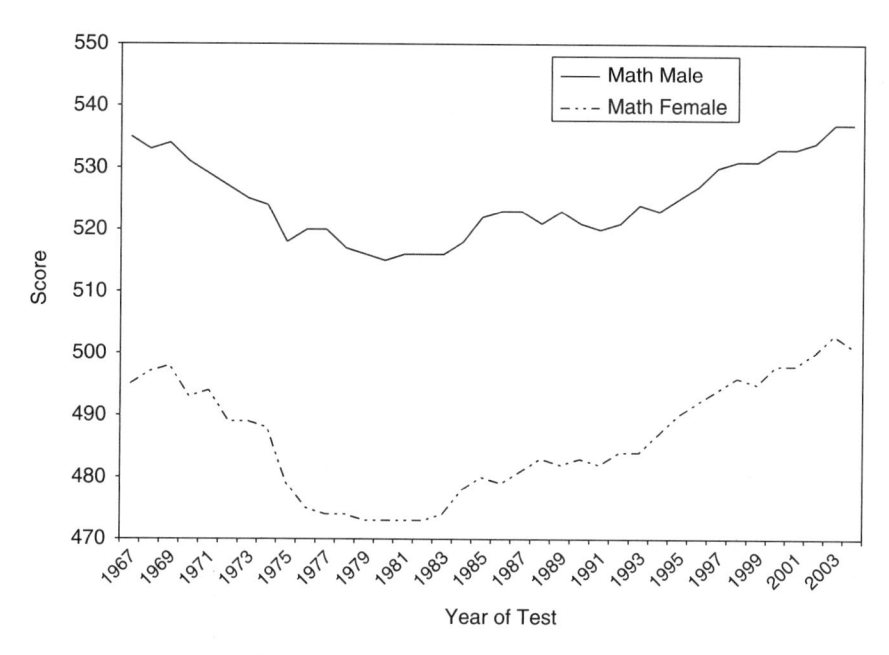

FIGURE 2-9 Average SAT scores of entering college classes, 1967-2004.
SOURCE: The College Board (2004). Table 2: Average SAT scores of entering college classes, 1967-2004. Date retrieved June 15, 2005, from http://www.collegeboard.com/prod_downloads/about/ness_info/cbsenior/yr2004/links.html.

mathematics courses, the male advantage on this test has remained fairly constant over the last 36 years.

Cognitive Process Taxonomy

How can we understand the grade-tests disparity? One way to consider the underlying cognitive processes used in executing the cognitive tasks being assessed when females or males excel at a cognitive task. Using a basic framework that was derived from the empirical literature on sex differences, Halpern (2000) proposed that females, in general, have faster access to information in episodic memory, to word knowledge and phonetic information; greater language fluency and implicit use of grammatical rules (in writing). Males, in general have faster access to visuospatial information and more accurate transformations of visuospatial information. In a study of the strategies used to solve mathematical problems, Gallagher et al. (2000) used the framework proposed by Halpern to see if boys and girls differed systematically in their use of mathematical strategies for different types of problems. In a series of several studies, they found that overall, the male students were more likely to use a flexible set of general strategies and more likely to solve problems that required a spatial representation, a short cut, or the maintenance of information in spatial working memory. Females were more likely to correctly solve problems with context that was familiar for females, used verbal skills, or required retrieval of a known solution or algebraic or multi-step solution.

Building on the cognitive processing model, Gallagher, Levin and Cahalan (2002) examined cognitive patterns of sex differences on math problems on the Graduate Record Examination (GRE). They found the same results as predicted from the processes involved in solving the specific math problems, with differences favoring males for problems where there was an advantage to using a spatially-based solution strategy (use of a spatial representation), but not when solution strategies were more verbal in nature or similar to the ones presented in popular math textbooks. Similarly, the usual male advantage was found with math problems that had multiple possible solution paths, but not on problems that had multiple steps, so the differences in the performance of males and females on GRE math problems lie in the recognition and/or selection of a solution strategy that may be novel and not in the load on working memory. They found that the usual male advantage on standardized math tests can be minimized, equated, or maximized by altering the way problems are presented and the type of cognitive processes that are optimal for their solution.

These are important findings because they advance our understanding of problem solving in general and math problem solving for all learners. These findings also suggest ways to help everyone improve at what is often the "funnel"—or sieve—in education. Everyone can be taught how to create spatial representations and how to use successful strategies when they are appropriate for a specific type of mathematical problem. This is one example where the study of sex differences can move us toward a better understanding of the cognitive processes people use and new ways to improve strategies for math problem solving.

Noncognitive Variables

There are many context variables that influence cognitive performance. The president of Harvard, Lawrence Summers (2005, January 14) offered a "high-powered job" hypothesis as one possible reason for the low participation rate of women at the full professor level in the sciences and math that considers the larger context of higher education. There are few women full professorships in any discipline at research universities—they are underrepresented in all disciplines. Higher education is one of the few places that has an early "up or out" system. Law and accounting firms that require early partnership are the only other comparable models where young talented employees must prove themselves in the first six or seven years of their careers. For a scientist, who will usually have a postdoc position after receiving a doctorate, tenure decisions will be made around age 36, which means that tenure clocks run in the same time zone as biological clocks. A recent study found that early babies—before tenure—hurt women's careers in academe, but help men's (Mason and Goulden, 2004). Women who want a career in academic science will have to make greater sacrifices than men, because in general, women have greater care responsibilities than men do. The inflexibility of the tenure system to accommodate to the reality of women's lives is the more likely and proximal cause of the underrepresentation of women in academic science, which in addition to the other requirements in the academy, includes long hours in the laboratory.

Thus, although there are sex differences in cognitive performance on many tests, and despite the many unanswered and important questions about the interplay of social, assessment, and biological variables on cognitive performance, the most immediate route to helping talented women gain entry and move through career in science and mathematics is by recognizing the family and other caretaking demands that most usually fall on women. Many talented women resent the choice between children and career that society is not asking of their male peers. Egalitarian households would go a long way to achieving workplace equity, but until we achieve that reality, part-time tenure track appointments without retaliation and other family-compatible options for men and women will be needed so that the nation can take advantage of the talent in the new workforce.

References

PL Ackerman, KR Bowen, ME Beier, and R Kanfer (2001). Determinants of individual differences and gender differences in knowledge. *Journal of Educational Psychology* 93:797-825.

MR Banaji and CD Hardin (1996). Automatic stereotyping. *Psychological Science* 7:136-141.

MJ Cullen, CM Hardison, and PR Sackett (2004). Using SAT-grade and ability-job performance relationships to test predictions derived from stereotype threat theory. *Journal of Applied Psychology* 89:220-239.

CA Dwyer and LM Johnson (1997). Grades, accomplishment, and correlates. In: *Gender and fair assessment*, Eds. WW Willingham and NS Cole, Mahwah, NJ: Erlbaum.

AM Gallagher, R De Lisi, PC Holst, AV McGillicuddy-De Lisi, M Morely, and C Cahalan (2000). Gender differences in advanced mathematical problem solving. *Journal of Experimental Child Psychology* 75:165-190.

A Gallagher, J Levin, and C Cahalan (2002). GRE Research: Cognitive patterns of gender differences in mathematics admissions test (ETS Report 02-19). Princeton, NJ: Educational Testing Service.

DF Halpern (2000). *Sex differences in cognitive abilities* (3rd ed.). Mahwah, NJ: Erlbaum.

DF Halpern (in press). Science, sex, and good sense. In: *Are Sex Differences in Cognition Responsible for the Underrepresentation of Women in Scientific Careers?* Eds. S Ceci and W Williams, Washington, DC: American Psychological Association.

DF Halpern and ML Collaer (2005). Sex differences in visuospatial abilities: More than meets the eye. In: *The Cambridge Handbook of Visuospatial Thinking*, Eds. P Shah and Miyake, Cambridge, MA: Cambridge University Press.

DF Halpern and U Tan (2001). Stereotypes and steroids: Using a psychobiosocial model to understand cognitive sex differences. *Brain and Cognition* 45:392-414.

LV Hedges and A Nowell (1995). Sex differences in mental test scores, variability, and numbers of high-scoring individuals. *Science* 269:41-45.

A Herlitz, L-G Nilsson, and L Baeckman (1997). Gender differences in episodic memory. *Memory and Cognition* 25:801-811.

JS Hyde (2005). The gender similarities hypothesis. *American Psychologist* 60:581-592.

MM Kimball (1989). A new perspective on women's math achievement. *Psychological Bulletin* 105:198-214.

AR Jensen (1998). *The g Factor: The Science of Mental Ability.* New York: Praeger.

LJ Levy, RS Astur, and KM Frick (2005). Men and women differ in object memory but not performance of a virtual radial maze. *Behavioral Neuroscience* 119:853-862.

R Lippa (1998). Gender-related individual differences and the structure of vocational interests of the people-things dimension. *Journal of Personality and Social Psychology* 74:996-1009.

EA Maguire, DG Gadian, IS Johnsrude, CD Ashburner, RSJ Frackowiak, and CD Frith (2000). Navigation-related structural change in the hippoccampi of taxi drivers. *Proceedings of National Academy of Sciences* 97:4398-4403.

MA Mason and M Goulden (2004). Do babies matter? The effect of family formation of the lifelong careers of academic men and women. *Academe.* Retrieved on-line March 28, 2005 from http://www.aaup.org.

MS Masters and B Sanders (1993). Is the gender difference in mental rotation disappearing? *Behavior Genetics* 23:337-341.

SD Moffat and E Hampson (1996). A curvilinear relationship between testosterone and spatial cognition in humans: Possible influence of hand preference. *Psychoneuroendocrinology* 21:323-337.

H Nordvik and B Amponsah (1998). Gender differences in spatial abilities and spatial activity among university students in an egalitarian educational system. *Sex Roles* 38:1009-1023.

MW O'Boyle, EJ Hoff, and HS Gill (1995). The influence of mirror reversals on male and female performance in spatial tasks: A componential look. *Personality and Individual Differences* 18:693-699.

M Peters, B Laeng, K Latham, M Jackson, R Zaiyouna, and C Richardson (1995). A redrawn Vandenberg and Kuse mental rotations test: Different versions and factors that affect performance. *Brain and Cognition* 28:39-58.

R Rosenthal, RL Rosnow, and DB Rubin, (2000). *Contrasts and effect sizes in behavioral research: A correlational approach.* Cambridge, UK: Cambridge University Press.

OC Schultheiss, MM Wirth, CM Torges, JS Pang, MA Villacota, and KM Welsh (2005). Effects of implicit power motivation on men's and women's implicit learning and testosterone changes after social victory or defeat. *Journal of Personality and Social Psychology* 88:174-188.

JN Shelton and JA Richseon. (2005). Intergroup contact and pluralistic ignorance. *Journal of Personality and Social Psychology* 88:91-107.

BB Sherwin (1999). Can estrogen keep you smart? Evidence from clinical studies. *Journal of Psychiatry and Neuroscience* 24:315-321.

LH Summers (2005, January 14). Remarks at NBER Conference on Diversifying the Science and Engineering Workforce, Cambridge, MA. Available online: http://www.president.harvard.edu/speeches/2005/nber.html. Retrieved: March 18, 2005.

U.S. Department of Education (1997). *The Third International Mathematics and Science Study.* Washington, DC: U.S. Department of Education. (http://www.ed.gov/nces.).

U.S. Department of Education (2000). National Center for Education Statistics, *Trends in Educational Equity for Girls and Women* NCES 2000-030, by Y Bae, S Choy, C Geddes, J Sable, and T Snyder. Washington, DC: U.S. Printing Office.

V Valian (1998). *Why So Slow? The Advancement of Women.* Cambridge, MA: The MIT Press.

MR Webb, D Lubinski, and CP Benbow (2002). Mathematically facile adolescents with math-science aspirations: New perspectives on their educational and vocational development. *Journal of Educational Psychology* 94:785-794.

J Wai, D Lubinski, and CP Benbow (2005). Creativity, and occupational accomplishments among intellectually precocious youths: An age 13 to 33 longitudinal study. *Journal of Educational Psychology* 97:484-492.

WOMEN IN SCIENCE AND MATHEMATICS: GENDER SIMILARITIES IN ABILITIES AND SOCIOCULTURAL FORCES *

Janet Shibley Hyde
Department of Psychology
University of Wisconsin

Abstract

Success in engineering and the physical sciences requires many abilities (Handelsman et al., 2005). Chief among them are mathematical, spatial, and verbal abilities, the first two for doing the science and the third for presenting one's work in scientific articles and at conferences. All three have been stereotyped as showing gender differences. Researchers have amassed mountains of data on gender differences in mathematical, spatial, and verbal abilities and have synthesized the finding using meta-analysis. This paper reviews these meta-analyses and other related research, concluding that gender differences in these abilities are generally small.

Success in engineering and the physical sciences requires many abilities (Handelsman et al., 2005). Chief among them are mathematical, spatial, and verbal abilities, the first two for doing the science and the third for presenting one's work in scientific articles and at conferences. All three have been stereotyped as showing gender differences. Researchers have amassed mountains of data on gender differences in mathematical, spatial, and verbal abilities. These data have been synthesized using a statistical technique called meta-analysis. Therefore, before reviewing the evidence on gender differences in abilities, I provide a brief explanation of meta-analysis.

Meta-Analysis

Meta-analysis is a statistical method for aggregating research findings across many studies of the same question (Hedges and Becker, 1986). It is ideal for synthesizing research on gender differences, an area in which often dozens or even hundreds of studies of a particular question have been conducted.

---

*Paper presented at the National Academies Convocation on Maximizing the Success of Women in Science and Engineering: Biological, Social, and Organizational Components of Success, held December 9, 2005, in Washington, DC. Preparation of this paper was supported in part by the National Science Foundation, Grant REC 0207109.

Crucial to meta-analysis is the concept of effect size, which measures the magnitude of the effect—in this case, the magnitude of the gender difference. In gender meta-analyses, the measure of effect size typically is d (Cohen, 1988).

$$d = \frac{M_M - M_F}{s_w}$$

where M_M is the mean score for males, M_F is the mean score for females, and s_w is the within-sex standard deviation. That is, d measures how far apart the male and female means are, in standardized units. In meta-analysis, the effect sizes computed from all individual studies are then averaged to obtain an overall effect size reflecting the magnitude of gender differences across all studies. Here I follow the convention that negative values of d mean that females scored higher and positive values of d indicate that males scored higher.

Although there is some disagreement among experts, a general guide is that an effect size d of 0.20 is a small difference, a d of 0.50 is moderate, and a d of 0.80 is a large difference (Cohen, 1988). As an example of a large effect, for the gender difference in throwing distance, $d = +1.98$ (Thomas and French, 1985).

Meta-analyses generally proceed in three steps: (1) The researchers locate all studies on the topic being reviewed, typically using databases such as PsychINFO and carefully chosen search terms. (2) Statistics are extracted from each report and an effect size is computed for each study. (3) An average of the effect sizes is computed to obtain an overall assessment of the direction and magnitude of the gender difference when all studies are combined.

Conclusions based on meta-analyses are almost always more powerful than conclusions based on an individual study, for two reasons. First, because meta-analysis aggregates over numerous studies, a meta-analysis typically represents the testing of tens of thousands—sometimes even millions—of participants. As such, the results should be far more reliable than those from any individual study. Second, findings from gender differences research are notoriously inconsistent across studies. For example, in the meta-analysis of gender differences in mathematics performance discussed later in this paper, 51% of the studies showed males scoring higher, 6% showed exactly no difference between males and females, and 43% showed females scoring higher (Hyde, Fennema, and Lamon, 1990). This makes it very easy to find a single study that supports one's prejudices. Meta-analysis overcomes this problem by synthesizing all available studies.

Gender Differences in Mathematics Performance

A major meta-analysis of studies of gender differences in mathematics performance surveyed 100 studies, representing the testing of more than 3 million persons (Hyde, Fennema, and Lamon, 1990). Averaged over all samples of the general population, $d = -0.05$, a negligible difference favoring females.

An independent meta-analysis confirmed the results of the first meta-analysis (Hedges and Nowell, 1995). It found effect sizes for gender differences in mathematics performance ranging between 0.03 and 0.26 across large samples of adolescents—all differences in the negligible to small range. Results from the International Assessment of Educational Progress also confirm that gender differences in mathematics performance are small across numerous countries including Hungary, Ireland, Israel, and Spain (Beller and Gafni, 1996).

For issues of the underrepresentation of women in the physical sciences, however, this broad assessment of the magnitude of gender differences is probably less useful than an analysis by both age and cognitive level tapped by the mathematics test. These results from one meta-analysis are shown in Table 2-1. Ages were grouped roughly into elementary school (ages 5-10 years), middle school (11-14), high school (15-18), and college age (19-25). Insufficient studies were available for older ages to compute mean effect sizes. Cognitive level of the test was coded as assessing either simple computation (requires the use of only memorized math facts, such as $7 \times 8 = 56$), conceptual (involves analysis or comprehension of mathematical ideas), problem solving (involves extending knowledge or applying it to new situations), or mixed. The results indicated that girls outperform boys by a small margin in computation in elementary school and middle school and there is no gender difference in high school. For understanding of mathematical concepts, there is no gender difference at any age level. For problem solving there is no gender difference in elementary or middle school, but a small gender difference favoring males emerges in high school and the college years. There are no gender differences, then, or girls perform better, in all areas except problem solving beginning in the high school years.

This gender difference in problem solving favoring males deserves attention because problem solving is essential to success in occupations in engineering and the physical sciences. Perhaps the best explanation for this gender difference, in

TABLE 2-1 The Magnitude of Gender Differences in Mathematics Performance as a Function of Age and Cognitive Level of the Test

Age group	Cognitive Level		
	Computation	Concepts	Problem solving
5-10	–0.20	–0.02	0.00
11-14	–0.22	–0.06	–0.02
15-18	0.00	0.07	0.29
19-25	NA	NA	0.32

SOURCE: Hyde et al. (1990).

view of the absence of a gender difference at earlier ages, is that it is a result of gender differences in course choice, i.e., the tendency of girls not to select optional advanced mathematics courses and science courses in high school. The failure to take advanced science courses may be particularly crucial because mathematics curricula often do not teach problem solving, whereas it typically is taught in chemistry and physics.

Gender Differences in Verbal Ability

A meta-analysis of studies of gender differences in verbal ability indicated that, overall, the difference was so small as to be negligible, $d = -0.11$ (Hyde and Linn, 1988). The negative value indicates better performance by females, but the magnitude of the difference is quite small. There are many aspects to verbal ability, of course. When analyzed according to type of verbal ability, the results were as follows: for vocabulary, $d = -0.02$; for reading comprehension $d = -0.03$; for speech production $d = -0.33$; and for essay writing $d = -0.09$. The gender difference in speech production favoring females is the largest and confirms females' better performance on measures of verbal fluency (not to be confused with measures of talking time). The remaining effects range from small to zero. Moreover, the magnitude of the effect was consistently small at all ages. Overall, then, gender difference in verbal ability are tiny and, if anything, favor females on measures such as essay writing and speech production, which should contribute to success in science. A second meta-analysis confirmed these findings using somewhat different methods (Hedges and Nowell, 1995).

Gender Differences in Spatial Ability

Spatial ability tests may tap any of several distinct skills: spatial visualization (finding a figure in a more complex one, like hidden-figures tests), spatial perception (identifying the true vertical or true horizontal when there is distracting information, such as the rod-and-frame task), and mental rotation (mentally rotating an object in 3 dimensions). Two meta-analyses are available on the question of gender differences in spatial performance. One found that the magnitude of gender differences varied substantially across the different types of spatial performance: $d = 0.13$ for spatial visualization, 0.44 for spatial perception, and 0.73 for mental rotation, all effects favoring males (Linn and Peterson, 1985). The last difference is large and potentially influential. The other meta-analysis found $d = 0.56$ for mental rotation (Voyer, Voyer, and Bryden, 1995), a somewhat smaller effect but nonetheless a substantial one. Gender differences in spatial performance—specifically, mental rotation—are important because mental rotation is crucial to success in several fields of engineering, chemistry, and physics (Hegarty and Sims, 1994).

Sociocultural Influences on Gender Differences in Mathematical and Spatial Abilities

The evidence on social and cultural influences on gender differences in mathematical and spatial abilities is plentiful and varied. I consider three categories of evidence: research on family and school influences, training studies, and cross-cultural analyses.

Family and School Influences

Abundant evidence exists for the multiple influences of parents and the schools on children's development. Here I focus on these influences specifically in the domains of abilities and academic performance. A limitation to some of these studies is that they report simply a correlation, for example, between parents' estimates of the child's mathematics ability and the child's score on a standardized test. From this correlation, we cannot infer the direction of causality with complete certainty. We cannot tell whether the parents' beliefs in the child influence the child's performance or whether the opposite process occurs—that children's test scores influence their parents' estimates of abilities. Moreover, it may be that both processes occur.

Numerous studies have confirmed the finding that parents' expectations for their children's academic abilities and success predict the children's self-concept of their own ability and their subsequent performance (e.g., Bleeker and Jacobs, 2004; Eccles, 1994). When engaged in a science task—playing with magnets—mothers talk about the science process (e.g., use explanations, generate hypotheses) more with boys than with girls (Tenenbaum et al., 2005). Moreover, the amount of mothers' science-process talk predicts children's comprehension of readings about science 2 years later. Observations of parents and children using interactive science exhibits at a museum showed that parents were three times more likely to explain science to boys than to girls (Crowley et al., 2001). Girls essentially grow up in a different family science environment than boys do.

Schools may exert their influence in multiple ways, including teachers' attitudes and behaviors, curriculum, ability grouping, and sex composition of the classroom. The availability of hands-on laboratory experiences is especially critical for learning in the physical sciences in middle school and high school. An important point is that, although laboratory experiences do not improve the physical science achievement of boys, they do improve the achievement of girls, thereby helping to close the gender gap in achievement in the physical sciences (Burkam, Lee, and Smerdon, 1997; Lee and Burkam, 1996). In science and mathematics classes, teachers are more likely to encourage boys than girls to ask questions and to explain (American Association of University Women, 1995; Jones and Wheatley, 1990; Kelly, 1988). In one study of high school geometry classrooms, teachers directed 61% of their praise comments to boys and 55% of their high-level open questions to boys (Becker, 1981). Experiences such as these are thought to give children a deeper conceptual knowledge of and more interest in science.

Students also exercise choice in school activities. Crucial to this discussion is their choice in high school to take advanced mathematics and science courses. The gender gap in mathematics course taking has narrowed over the last decade, so that by 1998 girls were as likely as boys to have taken advanced mathematics courses, including AP/IB calculus (National Science Foundation, 2005). Girls were actually slightly more likely than boys to take advanced biology (40.8% of girls, 33.8% of boys), AP biology (5.8% of girls, 5.0% of boys), and chemistry (59.2%, 53.3%). Boys, however, were more likely to take AP chemistry (3.3% of boys, 2.6% of girls) and physics (31% of boys, 26.6% of girls), and were twice as likely to take AP physics (2.3% of boys, 1.2% of girls) (National Science Foundation, 2005). The science pipeline heading toward physics, then, begins to leak early as fewer girls take the necessary high school courses to prepare themselves for college-level physics. It is beyond the scope of this article to review what psychologists know about the reasons why adolescents choose or do not choose to take challenging math and science courses. Readers wanting more information can look to a massive program of research conducted by Eccles (e.g., Eccles, 1994).

Training Studies

Environmental input is essential to the development of spatial and mathematical abilities (Baenninger and Newcombe, 1995; Newcombe, 2002; Spelke, 2005). Babies are not born knowing how to work calculus problems. Children acquire these skills through schooling and other experiences.

A meta-analysis found that spatial ability can indeed be improved with training, with effect sizes ranging between $d = 0.40$ to 0.80, depending on the length and specificity of the training (Baenninger and Newcombe, 1989). The effects of training were similar for males and females; that is, both groups benefited about equally from the training, and there was little evidence that the gender gap was closed or widened by training. A more recent study showed that the gender difference could be eliminated by carefully conceptualized training (Vasta et al., 1996). Unfortunately, most school curricula contain little or no emphasis on spatial learning. Girls, especially, could benefit greatly from such a curriculum.

The most recent development is multimedia software that provides training in 3-dimensional spatial visualization skills (Gerson, Sorby, Wysocki, and Baartmans, 2001). It has been used successfully with first-year engineering students. Most notably for the topic under discussion, there were improvements in the retention of women engineering students who took the spatial visualization course; without the course, the retention rate for women was 47%, whereas with the course it was 77%.

Cross-Cultural Analyses

The International Assessment of Educational Progress (IAEP) tested the math and science performance of 9- and 13-year-olds in 20 nations around the world. The effect sizes for gender differences for selected countries are shown in Table 2-2 (Beller and Gafni, 1996). Focusing first on the results for mathematics, we see that the gender differences are small in all cases. Most importantly, effect sizes are positive (favoring males) in some countries, negative (favoring females) in other countries, and several are essentially zero. The Trends in International Mathematics and Science Study (TIMSS, 2003, formerly the Third International Mathematics Study) found similar results, with some positive and some negative effect sizes, and most < 0.10. In the TIMSS data for eighth graders, the magnitude of the gender difference was 0.09 in Chile (country average score 379), 0.02 in the United States (country average 502), 0.01 in Japan (country average 569), and –0.05 in Singapore (country average 611). That not only the magnitude, but also the direction of gender differences in mathematics performance varies from country to country is powerful testimony to the importance of sociocultural factors in shaping those differences. Perhaps most importantly, though, the gender difference is very small in most nations.

Focusing next on the results for science performance (Table 2-2), we can see that the effect sizes more consistently favor males and are somewhat larger, although not large for any nation. When the results are broken down by science, gender differences are smaller in life sciences knowledge (0.11 and 0.20 at ages 9 and 13, respectively, averaged over all countries) and somewhat larger for physical sciences (0.22 and 0.33) (Beller and Gafni, 1996).

TABLE 2-2 Effect Sizes for Gender Differences in Mathematics and Science Test Performance Across Countries

| | Mathematics | | Science | |
Country	9 years	13 years	9 years	13 years
Hungary	–0.03	–0.02	0.09	0.25
Ireland	–0.06	0.19	0.20	0.31
Israel	0.16	0.15	0.23	0.24
Korea	0.28	0.10	0.39	0.31
Scotland	–0.01	–0.02	–0.01	0.20
Spain	0.01	0.18	0.25	0.24
Taiwan	0.03	0.02	0.25	0.08
U. S.	0.05	0.04	0.09	0.29
All countries	0.04	0.12	0.16	0.26

SOURCE: Beller and Gafni, 1996, Table 2 and Appendix.

It is important to note that cross-cultural differences in mathematics performance are enormous compared with gender differences in any one country. For example, in one cross-national study of 5th graders, American boys (M = 13.1) performed better than American girls (M = 12.4) on word problems, but 5th grade Taiwanese girls (M = 16.1) and Japanese girls (M = 18.1) performed far better than American boys (Lummis and Stevenson, 1990). Culture is considerably more important than gender in determining mathematics performance.

In perhaps the most sophisticated analysis of cross-national patterns of gender differences in mathematics performance, the researchers found that, across nations, the magnitude of the gender difference in mathematics performance for eighth graders correlated significantly with a variety of measures of gender stratification in the countries (Baker and Jones, 1993). For example, the magnitude of the gender difference in math performance correlated –0.55, across nations, with the percentage of women in the workforce in those nations. That is, the more that women participate in the labor force (an index of gender equality), the smaller the gender difference in mathematics achievement.

The Gender Similarities Hypothesis

I propose an alternative to our cultural and scientific obsession with gender differences. The alternative is the Gender Similarities Hypothesis, which I formalized in an article that appeared in the *American Psychologist* this year (Hyde, 2005). For that paper, I essentially meta-analyzed meta-analyses. That is, I found all the meta-analyses of psychological gender differences that I could. I found 46 relevant meta-analyses, and from them I extracted 124 effect sizes—d's—for gender differences. The meta-analyses spanned a wide range of psychological characteristics, including abilities, communication, aggression, leadership, personality, and self-esteem.

I organized those 124 effect sizes into ranges— those that are close to zero, i.e., in the range 0 to 0.10, those that are small, 0.11 to 0.35, those that are moderate in magnitude, 0.36 to 0.65, those that are large, 0.66 to 1.00, and those that are very large, > 1.00. The results indicated that 30% of those effect sizes were in the close-to-zero range, and another 48% were small. So, 78% of the effect sizes were small or close to zero—that is the gender similarities hypothesis—psychologically, women and men are more similar than they are different. There are a few exceptions of large differences, but the big picture is one of gender similarities.

Implications: How Can We Close the Gender Gap in Engineering and the Physical Sciences?

One conclusion of this review is that, overall, there are no gender differences in math performance, but a gender difference favoring males in complex problem solving does emerge in high school. Mathematical problem solving is crucial to success in the physical sciences, so this gap must be addressed. The evidence also indicates a gender gap in favor of males in spatial ability, specifically in mental rotation. This ability, too, is crucial to success in the physical sciences and must be addressed.

The following policy recommendations flow from the data reviewed here:

1. Focusing on the gender difference in spatial skill, we need to institute a spatial learning curriculum in the schools. Girls are seriously disadvantaged by its absence.
2. Colleges of engineering should have a spatial skills training program for entering students. Theoretically, such a program should help in physics and chemistry as well.
3. We should *require* 4 years of math and 4 years of science in high school— or at least require it for university admission. Otherwise, girls will elect not to take some advanced science courses and, without carefully making the decision, close themselves out of outstanding careers in engineering and the sciences.
4. The mathematics curriculum in many states continues to need attention. It needs far more emphasis on real problem solving, and that approach will benefit not only girls, but boys as well.
5. Hands-on science labs will benefit girls and help to close the gender gap. And, they represent good science education practice.
6. Teachers and high-school guidance counselors need to be educated about the findings on gender similarities in math performance. Otherwise, teachers will believe the stereotypes about girls' math inferiority that pervade our culture, the teachers will have lower expectations for girls' math performance, and those expectations will convey themselves to the students.

If we do all this—and much more—we can all look forward to a day when girls and women will have equal access to careers in engineering and the sciences. And our nation will benefit from maximizing women's contributions.

References

American Association of University Women (1995). *How schools shortchange girls*. Washington, DC: AAUW.

M Baenninger and N Newcombe (1989). The role of experience in spatial test performance: A meta-analysis. *Sex Roles* 20:327-344.

M Baenninger and N Newcombe (1995). Environmental input to the development of sex-related differences in spatial and mathematical ability. *Learning and Individual Differences* 7:363-379.

DP Baker and DP Jones (1993). Creating gender equality: Cross-national gender stratification and mathematical performance. *Sociology of Education* 66:91-103.

JR Becker (1981). Differential treatment of females and males in mathematics classes. *Journal for Research in Mathematics Education* 12:40-53.

M Beller and N Gafni (1996). The 1991 International Assessment of Educational Progress in Mathematics and Sciences: The gender differences perspective. *Journal of Educational Psychology* 88:365-377.

MM Bleeker and JE Jacobs (2004). Achievement in math and science: Do mothers' beliefs matter 12 years later? *Journal of Educational Psychology* 96:97-109.

DT Burkam, VE Lee, and BA Smerdon (1997). Gender and science learning early in high school: Subject matter and laboratory experiences. *American Educational Research Journal* 34:297-331.

J Cohen (1988). *Statistical power analysis for the behavioral sciences.* 2nd ed., Hillsdale, NJ: Erlbaum.

K Crowley, MA Callanan, HR Tenenbaum, and E Allen (2001). Parents explain more often to boys than to girls during shared scientific thinking. *Psychological Science* 12:258-261.

JS Eccles (1994). Understanding women's educational and occupational choices: Applying the Eccles et al. model of achievement-related choices. *Psychology of Women Quarterly* 18:585-610.

H Gerson, SA Sorby, A Wysocki, and BJ Baartmans (2001). The development and assessment of multimedia software for improving 3-D spatial visualization skills. *Computer Applications in Engineering Education* 9:105-113.

J Handelsman, N Cantor, M Carnes, D Denton, E Fine, B Grosz, V Hinshaw, C Marrett, S Rosser, D Shalala, and J Sheridan (2005). More women in science. *Science* 309:1190-1191.

LV Hedges and BJ Becker (1986). Statistical methods in the meta-analysis of research on gender differences. In J. S. Hyde & M. C. Linn (Eds.), *The psychology of gender: Advances through meta-analysis* (pp. 14-50). Baltimore: Johns Hopkins University Press.

LV Hedges and A Nowell (1995). Sex differences in mental test scores, variability, and numbers of high-scoring individuals. *Science* 269:41-45.

M Hegarty and VK Sims (1994). Individual differences in mental animation during mechanical reasoning. *Memory and Cognition* 22(4):411-430.

JS Hyde (2005). The gender similarities hypothesis. *American Psychologist* 60:581-592.

JS Hyde, E Fennema, and SJ Lamon (1990). Gender differences in mathematics performance: A meta-analysis. *Psychological Bulletin* 107:139-155.

JS Hyde and MC Linn (1988). Gender differences in verbal ability: A meta-analysis. *Psychological Bulletin* 104:53-69.

MG Jones and J Wheatley (1990). Gender differences in teacher-student interactions in science classrooms. *Journal of Research in Science Teaching* 27:861-874.

A Kelly (1988). Gender differences in teacher-pupil interactions: A meta-analytic review. *Research in Education* 39:1-23.

VE Lee and DT Burkam (1996). Gender differences in middle grade science achievement: Subject domain, ability level, and course emphasis. *Science Education* 80:613-650.

MC Linn and AC Petersen (1985). Emergence and characterization of sex differences in spatial ability: A meta-analysis. *Child Development* 56:1479-1498.

M Lummis and HW Stevenson (1990). Gender differences in beliefs and achievement: A cross-cultural study. *Developmental Psychology* 26:254-263.

National Science Foundation (2002). *Science and engineering degrees, 1966-2000* (NSF 02-327), Author Susan T. Hill. Arlington, VA: National Science Foundation. http://ww.nsf.gov/sbe/srs/stats.htm.

National Science Foundation (2005). *Science and engineering indicators 2004: Elementary and secondary education, mathematics and science coursework and student achievement.* Arlington, VA: National Science Foundation. http://www.nsf.gov/statistics/seind04/.

NS Newcombe (2002). The nativist-empiricist controversy in the context of recent research on spatial and quantitative development. *Psychological Science* 13:395-401.

E Spelke (2005). Sex differences in intrinsic aptitude for mathematics and science? A critical review. *American Psychologist* 60(9):950-958.

HR Tenenbaum, CE Snow, KA Roach, and B Kurland (2005). Talking and reading science: Longitudinal data on sex differences in mother-child conversations in low-income families. *Applied Developmental Psychology* 26:1-19.

JK Thomas and KE French (1985). Gender differences across age in motor performance: A meta-analysis. *Psychological Bulletin* 98:260-282.

Trends in International Mathematics and Science Study (TIMSS) (2003). http://nces.ed.gov/pubs2005/timss03.

R Vasta, JA Knott, and CE Gaze (1996). Can spatial training erase the gender differences on the water-level task? *Psychology of Women Quarterly* 20:549-568.

D Voyer, S Voyer, and MP Bryden, (1995). Magnitude of sex differences in spatial abilities: A meta-analysis and consideration of critical variables. *Psychological Bulletin* 117:250-270.

CREATING AN INCLUSIVE WORK ENVIRONMENT*

Sue V. Rosser
Ivan Allen School of Liberal Arts and Technology
Georgia Institute of Technology

Abstract

Faced with a severe shortage of scientists and engineers, exacerbated by changes in immigration policies in the wake of 9/11, the United States has renewed its efforts to diversify the scientific and technological workforce, including attracting and retaining women in academic science and engineering. At the dawn of the 21st century, several promising developments, particularly the National Science Foundation's ADVANCE program, indicate the willingness of the scientific and engineering professions and the academy to address the under-representation of women in academic ranks that has continued for decades, despite federally and foundation-funded programs to increase the number of female faculty members (Rosser and Lane, 2002).

In March 1999 the Massachusetts Institute of Technology released "A Study on the Status of Women Faculty in Science at MIT"[1] creating a stir that spread far beyond the institutional boundaries of MIT. More than one year later, MIT President Charles Vest hosted a meeting of the presidents, chancellors, provosts, and twenty-five women scientists from some of the most prestigious research univer-

---

*Paper presented at the National Academies Convocation on Maximizing the Success of Women in Science and Engineering: Biological, Social, and Organizational Components of Success, held December 9, 2005, in Washington, DC.

[1]Published in the *MIT Faculty Newsletter* XI (4). Available at http://web.mit.edu/fnl/women/women.html.

sities in the country. At the close of the meeting on January 29, 2001, they issued a joint statement recognizing "that this challenge will require significant review of, and potentially significant change in, the procedures within each university, and within the scientific and engineering establishments as a whole." (Campbell, 2001), thus acknowledging that institutional barriers have prevented women scientists and engineers from having a level playing field and that science and engineering might need to change to accommodate women.

Almost simultaneously, the National Science Foundation (NSF) initiated ADVANCE, a new awards program that provided funding of $17 million for fiscal year 2001. The program offers an award for institutional, rather than individual solutions to empower women to participate fully in science and technology. NSF encouraged institutional solutions because of "increasing recognition that the lack of women's full participation at the senior level of academe is often a systemic consequence of academic culture" (NSF, 2001a). Under ADVANCE, NSF grants Institutional Transformation Awards, ranging up to $750,000 per year for up to five years, to promote the increased full participation and advancement of women; Leadership Awards recognize the work of outstanding organizations of individuals and enable them to sustain, intensify and initiate new activity (NSF, 2001a).

ADVANCE Institutions

In October, 2001 the first eight institutions receiving ADVANCE awards were announced (NSF, 2001b): Georgia Tech, New Mexico State, the University of California-Irvine, the University of Colorado-Boulder, the University of Michigan, the University of Puerto Rico, the University of Washington, and the University of Wisconsin-Madison. Hunter College joined the first round of ADVANCE awardee institutions in early 2002.

In 2003, NSF announced 10 second round institutional transformation grants: Case Western Reserve, Columbia University, Kansas State University, University of Alabama-Birmingham, University of Maryland-Baltimore County, University of Montana, University of Rhode Island, University of Texas-El Paso, Utah State, and Virginia Tech. The third round of ADVANCE institutional proposals should be announced early in 2006.

ADVANCE promises to go beyond individual research projects of women scientists and engineers that previous NSF initiatives such as Professional Opportunities for Women in Research and Education (POWRE), Faculty Awards for Women (FAW), Career Advancement Awards (CAA), and Visiting Professorships for Women (VPW) (Rosser & Lane, 2002) supported to solve problems with broader systemic and institutional roots such as balancing career and family.

Institutional Initiatives

The NSF-funded ADVANCE initiative and the MIT Report have brought attention to the need for institutional transformation to improve the daily lives of all faculty, particularly women scientists and engineers. Other institutions have undertaken transformation initiatives using their own funds. In the wake of issues raised by the comments made by President Summers on January 14, 2005, Harvard announced a $50 million initiative on May 16, 2005 in an attempt to address these issues (Pope, 2005). Princeton has undertaken several efforts, including an automatic one year family leave extension on the tenure track for both men and women, rather than placing the onus on the faculty member to ask for the extension (Bartlett, 2005).

Priorities for Institutional Change: Lessons from POWRE

To be most effective, proposed institutional changes should address the institutional barriers identified as most problematic by women scientists and engineers. Data from the almost 400 respondents to an e-mail survey of fiscal years 1997, 1998, 1999, and 2000 NSF Professional Opportunities for Women in Research and Education (POWRE) awardees reveal the barriers academic women scientists and engineers identify as most challenging to their careers. POWRE awardees were women who received peer-reviewed funding from a focused National Science Foundation program (NSF, 1997) from fiscal years (FY) 1997-2000. Because POWRE was the NSF initiative that ADVANCE replaced in 2001, the quantitative and qualitative data from the entire POWRE awardee cohort are particularly relevant in exposing the barriers that institutions should change to empower and enable women scientists and engineers.

Women scientists and engineers who were U.S. citizens at any rank in tenured, tenure track, or nontenure track positions at any four-year college, comprehensive, or research university were eligible to apply to POWRE. Although a few tenured full professors, faculty from four-year institutions, and/or nontenure track individuals received awards, the vast majority of POWRE awardees were untenured assistant professors in tenure track positions at research universities.

All POWRE new grant awardees were sent questionnaires via e-mail that included the question "What are the most significant issues/challenges/opportunities facing women scientists today as they plan their careers?" Overwhelming numbers of respondents across all 4 years found "balancing work with family" (Response 1) to be the most significant challenge facing women scientists and engineers (Table 2-3). When analyzed by disciplines, the responses of women remained remarkably similar across the disciplines, with balancing work with family responsibilities as the major issue for women from all disciplines (Rosser and Lane, 2002).

Table 2-4 groups the responses to Question 1 into four categories. Adding

TABLE 2-3 Total Responses to Question 1

Question 1: What are the most significant issues/challenges/opportunities facing women scientists today as they plan their careers?

Categories	1997 % of responses		1998 % of responses		1999 % of responses		2000 % of responses	
1 Balancing work with family responsibilities (children, elderly relatives, etc.)	62.7	(42/67)	72.3	(86/119)	77.6	(76/98)	71.4	(75/105)
2 Time management/balancing committee responsibilities with research and teaching	22.4	(15/67)	10.1	(12/119)	13.3	(13/98)	13.3	(14/105)
3 Low numbers of women, isolation and lack of camaraderie/mentoring	23.9	(16/67)	18.5	(22/119)	18.4	(18/98)	30.5	(33/105)
4 Gaining credibility/respectability from peers and administrators	22.4	(15/67)	17.6	(21/119)	19.4	(19/98)	21.9	(23/105)
5 "Two career" problem (balance with spouse's career)	23.9	(16/67)	10.9	(13/119)	20.4	(20/98)	20.0	(21/105)
6 Lack of funding/inability to get funding	7.5	(5/67)	4.2	(5/119)	10.2	(10/98)	8.6	(9/105)
7 Job restrictions (location, salaries, etc.)	9.0	(6/67)	9.2	(11/119)	7.1	(7/98)	5.7	(6/105)
8 Networking	6.0	(4/67)	<1	(1/119)	0	(0/98)	4.8	(5/105)
9 Affirmative action backlash/discrimination	6.0	(4/67)	15.1	(18/119)	14.3	(14/98)	12.4	(13/105)
10 Positive: active recruitment of women/more opportunities	6.0	(4/67)	10.1	(12/119)	9.2	(9/98)	14.3	(15/105)
11 Establishing independence	3.0	(2/67)	0	(0/119)	6.1	(6/98)	2.9	(3/105)
12 Negative social images	3.0	(2/67)	3.4	(4/119)	2.0	(2/98)	<1	(1/105)
13 Trouble gaining access to nonacademic positions	1.5	(1/67)	1.7	(2/119)	1.0	(1/98)	1.9	(2/105)
14 Sexual harassment	1.5	(1/67)	<1	(1/119)	2.0	(2/98)	1.9	(2/105)
15 No answer	0	(0/67)	<1	(1/119)	1.0	(1/98)	1.9	(2/105)
16 Cut-throat competition	—	—	—	—	1.0	(1/98)	1.9	(2/105)

TABLE 2-4 Categorization of Question 1 Across Year of Award

Question 1: What are the most significant issues/challenges/opportunities facing women scientists today as they plan their careers?

Categories	Response numbers[b]	Means of responses 1997	1998	1999	2000
A Pressures women face in balancing career and family	1, 5, 7	31.9%	30.8%	35.0%	32.4%
B[a] Problems faced by women because of their low numbers and stereotypes held by others regarding gender	3, 4, 8, 10, 12	12.3%	10.1%	9.8%	14.5%
C[a] Issues faced by both men and women scientists and engineers in the current environment of tight resources, which may pose particular difficulties for women	2, 6, 16	10.0%	4.8%	8.2%	7.9%
D More overt discrimination and harassment	9, 11, 13, 14	3.0%	4.4%	5.8%	4.8%

[a]The alphabetic designation for categories B and C have been exchanged, compared with earlier papers (Rosser and Zieseniss, 2000) to present descending response percentages.

[b]Given the responses from all four years, after receiving faculty comments at various presentations of this research, and after working with the data, we exchanged two questions from both category B and D to better reflect the response groupings. Specifically, responses 10 and 12 (considered in category D in Rosser and Zieseniss, 2000) were moved to category B. Similarly, responses 11 and 13 (included in category B in Rosser and Zieseniss, 2000) were placed into category D.

restrictions because of spousal situations (Responses 5 and 7) to "balancing work with family responsibilities" (Response 1) suggests that Category A-pressures women face in balancing career and family is the most significant barrier identified by women scientists and engineers. A second grouping (Responses 3, 4, 8, 10, and 12) appears to result from the low numbers of women scientists and engineers and consequent stereotypes that surround expectations about their performance. Isolation and lack of mentoring as well as gaining credibility and respectability from peers and administrators typify Category B. Category C (Responses 2, 6, 16) includes issues men and women scientists and engineers face in the current environment of tight resources that may pose particular difficulties for women because of their low numbers or their balancing act between career and family. For example, time management issues such as balancing committee responsibilities with research and teaching (Response 2) can be a problem for male as well as female faculty. However, because of their low numbers in science and engineering, women faculty are often asked to serve on more committees to meet gender diversity needs, even while they are still junior, and to advise more students, either formally or informally (Rosser and Zieseniss, 2000). Cut-throat competition makes it difficult for both men and women to succeed and obtain funding. Gender stereotypes that reinforce women's socialization to be less overtly competitive may make it more difficult for a woman scientist or engineer to succeed in a very competitive environment. Category D (Responses 9, 11, 13, 14) identifies barriers of overt harassment and discrimination women scientists and engineers face. Sometimes even a positive response, such as active recruitment of women (Response 10) leads to backlash and difficulty gaining credibility from peers who assume a woman obtained her position because of affirmative action.

Example quotations from the respondents from all 4 years provide the qualitative context for the categories:

Category A: Pressures Women Face in Balancing Career and Family

- "At the risk of stereotyping, I think that women generally struggle more with the daily pull of raising a family or caring for elderly parents, and this obviously puts additional demands on their time. This is true for younger women, who may struggle over the timing of having and raising children, particularly in light of a ticking tenure clock, but also for more senior women, who may be called upon to help aging parents (their own or in-laws). Invariably they manage, but not without guilt." (2000 respondent 63)
- "Managing dual-career families (particularly dual academic careers). Often women take the lesser position in such a situation. PhD women are often married to PhD men. Most PhD men are not married to PhD women." (2000 respondent 16)

Category B: Problems Faced by Women Because of Their Low Numbers and Stereotypes Held by Others Regarding Gender

- "The biggest challenge that women face in planning a career in science is not being taken seriously. Often women have to go farther, work harder, and accomplish more in order to be recognized." (2000 respondent 21)
- "In my field (concrete technology), women are so poorly represented that being female certainly creates more notice for you and your work, particularly when presenting at conferences. This can be beneficial, as recognition of your research by your peers is important for gaining tenure; it can also add to the already large amount of pressure on new faculty." (2000 respondent 70)

Category C: Issues Faced by Men and Women Scientists and Engineers in the Current Environment of Tight Resources, Which May Pose Particular Difficulties for Women

- "I have noticed some problems in particular institutions I have visited (or worked at) where women were scarce. As a single woman, I have sometimes been viewed as 'available,' rather than as a professional co-worker. That can be really, really irritating. I assume that single men working in a location where male workers are scarce can face similar problems. In physics and astronomy, usually the women are more scarce." (1997 respondent 26)
- "I still find the strong perception that women should be doing more teaching and service because of the expectation that women are more nurturing. Although research as a priority for women is given a lot of lip service, I've not seen a lot of support for it." (2000 respondent 1)

Category D: More Overt Discrimination and/or Harassment

- "There are almost no women in my field, no senior women, and open harassment and discrimination are very well accepted and have never been discouraged in any instance I am aware of." (1998 respondent 53)
- "I have often buffered the bad behavior of my colleagues—and over the years I have handled a number of sexual harassment or 'hostile supervision' cases where a more senior person (all of them male) was behaving inappropriately toward a lower social status woman (or in rarer cases a gay man)." (1999 respondent 59)

Models for best practices

Considerable research has revealed barriers such as balancing career and family, dual career issues, isolation, dearth of mentoring, and possible unconscious bias in search processes, negotiation, evaluation, as well as promotion and tenure, that may differentially impact appointment, retention, and advancement of women faculty. Coupling this research with evidence from climate surveys and experiences raised by faculty at their own institutions, models for some best practices have begun to emerge from ADVANCE institutions.

At the time of this conference, none of the ADVANCE institutions had completed their five-year institutional transformation grant, so evaluation of the success of these programs is not possible. However, progress towards goals may be gleaned from examination of the reports submitted annually to NSF on the projects (http://www.nsf.gov/advance) and from many of the ADVANCE institution websites. Although most of the efforts have centered on advancing junior women to senior ranks, anecdotal evidence and some preliminary data from my current research suggest that senior women scientists and engineers face different obstacles. Institutions need also to address these barriers to retain senior women and insure they reach their full potential. Preliminary data from the Georgia Tech ADVANCE project grant, which ends in fall 2006, suggest that more women have been promoted to full professor, endowed chairs, and administrative positions since the grant was obtained in October, 2001. See Figures 2-10, 2-11A, and 2-11B.

Family friendly policies and practices

To facilitate the balancing of career and family, perceived overwhelmingly by women scientists and engineers, particularly those of younger ages, as the major issues (Rosser, 2004), Georgia Tech instituted the following family friendly policies and practices: stop the tenure clock; active service, modified duties; lactation stations; and day care.[2] Many other institutions have similar and additional policies, including for flexible work hours at Johns Hopkins[3] and the University of Wisconsin,[4] and shared positions for dual career couples at Cornell[5] and Grinnell.[6]

---

[2]The specific details of these policies can be accessed under Family and Work Policies at http://www.advance.gatech.edu. Retrieved on June 23, 2005.

[3]http://hrnt.jhu.edu/worklife/benefits/flex/index.cfm. Retrieved on November 21, 2005.

[4]http://www.secfac.wisc.edu/governance/legistlaiton/Pages300-399.htm#308. Retrieved on November 21, 2005.

[5]http://www.policy.cornell.edu/PDF_6613_Workplace_Flexibility.cfm. Retrieved on November 21, 2005.

[6]http://www.grinnell.edu/offices/dean/chairinfo/sharedpos/. Retrieved on November 21, 2005.

SECTION 2: SELECTED WORKSHOP PAPERS

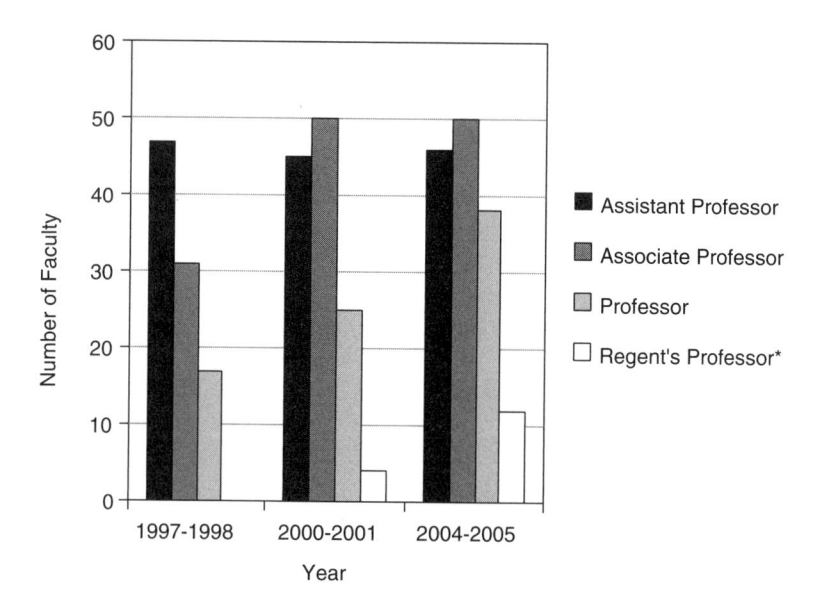

FIGURE 2-10 Georgia Institute of Technology female faculty by rank and year, institution-wide.
* Regent's Professorships are a rare, distinguished promotion above the level of full professor, which are open to both men and women faculty at the research institutions in the University System of Georgia. A Regent's Professorship is awarded only upon the unanimous recommendation of the president, the dean of the graduate school, the administrative dean, the academic dean, and three other members of the faculty to be named by the president, and upon the approval of the Chancellor and the Committee on Education.

Speed-mentoring

To assist junior faculty in preparation for tenure and/or promotion, Georgia Tech ADVANCE Professor Jane Ammons developed a workshop in which junior faculty members consult for 15 to 20 minutes with each of four experienced tenure case reviewers who identify gaps and offer suggestions for strengthening the tenure case. Even more women seeking promotion to full professor than tenure and promotion to associate professor attended the workshop, confirming information revealed in the climate survey that individuals understand the parameters less well for promotion to full professor.

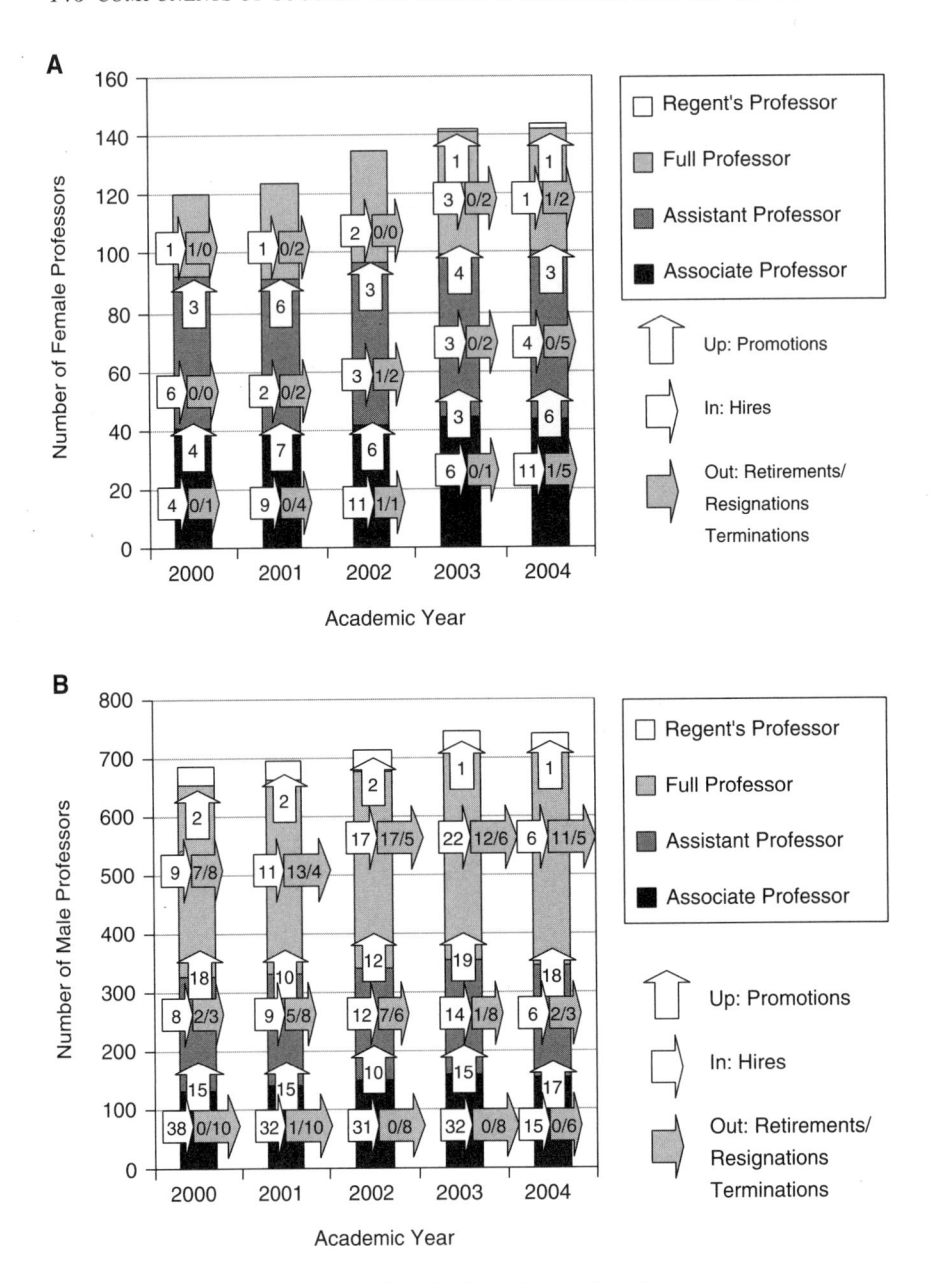

146 COMPONENTS OF SUCCESS FOR WOMEN IN ACADEMIC SCIENCE & ENGINEERING

FIGURE 2-11 Georgia Institute of Technology faculty flux charts.
A: Female Faculty; B: Male Faculty

Training of search committees

The University of Wisconsin-Madison has designed workshops to train search committees in good search methods, including sensitization to bias.[7] This training includes cultivation of professional relationships with promising women scholars at professional meetings, active solicitation of applications from qualified women, and deliberate actions to overcome unconscious bias such as encouraging time for thorough review and evaluation of each individual to insure focus on data rather than impressions. Denice Denton, while at the University of Washington, developed a Faculty Recruitment Toolkit.[8] The University of Michigan developed the STRIDE program, led by faculty to improve diversity and excellence in recruiting.[9]

Training of chairs and deans

Because top administrators can set the climate and standards for fostering inclusivity, programs to train department chairs to recognize and combat isolation, while nurturing inclusion become critical. The ADVANCE program at the University of Michigan worked with an interactive theater program that portrays typical academic situations and engages academic audiences in discussion around interpersonal behaviors affecting these issues. The University of Washington has developed a National UW ADVANCE Summer Leadership Workshop for Department Chairs.[10]

Training of tenure and promotion committees

To minimize gender, racial, and other biases in promotion and tenure, the Provost at Georgia Tech, who also serves as Principal Investigator on its ADVANCE grant, appointed a Promotion and Tenure ADVANCE Committee (PTAC) to assess existing promotion and tenure processes, explore potential forms of bias, provide recommendations to mitigate against them, and elevate awareness of both candidates and committees for expectations and best practices in tenure and promotion. After one year of study, the committee developed nine case studies with accompanying sample curriculum vitae that served as the basis for an interactive Web-based instrument. This interactive Web tool, Awareness

---

[7] http://wiseli.engr.wisc.edu/products.htm. Retrieved on November 21, 2005.
[8] http://www.washington.edu/admin/eoo/forms/ftk_01.html. Retrieved on November 21, 2005.
[9] http://www.umich.edu/%7Eadvproj/stridepresents_files/fram.htm. Retrieved on November 21, 2005.
[10] http://www.engr.washington.edu/advance/workshops/National/Workshop/chair-workshop.html. Retrieved on November 21, 2005.

of Decision in Evaluation of Promotion and Tenure (ADEPT), is designed to allow individuals to participate in a virtual promotion and tenure meeting.[11]

Each ADVANCE institution has evolved programs and policies to address similar issues on its campus. Most have at least one program that is unique, which if successful, might serve as a model for other institutions. Virginia Tech hosts the ADVANCE portal website for all ADVANCE institutional transformation awardees.[12] As these models spread to other campuses where they undergo implementation and improvements, a national transformation of science and engineering may occur that fulfills the promise of the Science and Technology Equal Opportunities Act to create a scientific and technological community reflective of our diverse society.

References

T Bartlett (2005). More time. *The Chronicle of Higher Education* 52(2):A-16.

K Campbell (2001). Leaders of 9 universities and 25 women faculty meet at MIT, agree to equity reviews. *MIT News Office.* Available at http://web.mit.edu/newsoffice/nr/2001/gender.html.

National Science Foundation. (1997). *Professional opportunities for women in research and education. Program announcement.* Arlington, VA: National Science Foundation.

National Science Foundation. (2001a). *ADVANCE. Program solicitation.* Arlington, VA: National Science Foundation.

National Science Foundation. (2001b). *ADVANCE Institutional Transformation Awards.* http://www.nsf.gov/advance. Retrieved on October 1, 2001.

J Pope (2005). Harvard to commit $50M to women's programs. *The Boston Globe* (17 May). http://www.boston.com/news/education/higher/articles/2005/05/17/html.

SV Rosser 2004. *The Science Glass Ceiling: Academic Women Scientists and the Struggle to Succeed.* New York: Routledge.

SV Rosser and EO Lane (2002). Key barriers for academic institutions seeking to retain women scientists and engineers: Family-unfriendly policies, low numbers, stereotypes, and harassment. *Journal of Women and Minorities in Science and Engineering* 8(2):161-190.

SV Rosser and M Zieseniss (2000). *Final report on professional opportunities for women in research and education (POWRE) workshop.* Gainesville, FL: Center for Women's Studies and Gender Research.

---

[11]The Web-based instrument, along with best practices from PTAC, and resources on bias can be accessed at http://www.adept.gatech.edu/ptac. Retrieved on June 23, 2005.

[12]The portal can be accessed at http://www.advance.vt.edu. Retrieved on June 23, 2005.

LONG TIME NO SEE: WHY ARE THERE STILL SO FEW WOMEN IN ACADEMIC SCIENCE AND ENGINEERING?*

Joan C. Williams
UC Hastings College of the Law
Center for WorkLife Law

Abstract

After all these years, all these reports, all these initiatives, why does the percentage of women academics in science and engineering remain so low? The traditional response is to point to the "chilly climate" for women. That metaphor is outdated. What keeps women back is gender bias, although it does "not look like what we thought discrimination looked like." The time has come to link the chilly climate with two literatures that have flowered since the "climate" metaphor was invented in 1982 by Roberta Hall and Bernice Sandler (Sandler et al., 1996; Sullivan, 2005).

The first is the growing literature in experimental social psychology on stereotyping and cognitive bias, which shows that many of the patterns that create a "built-in headwinds" for women in the sciences and engineering reflect documented patterns of gender bias (Griggs v. Duke Power Co., 401 U.S. 424 (1971)). The second is antidiscrimination law, which increasingly accepts stereotyping evidence in court and highlights that "chilly climate" patterns may be illegal. This article provides a very brief introduction to both literatures. Before it does so, it provides an even briefer introduction to a third discipline that provides crucial data for understanding why women's progress has been so glacially slow: demography (Valian, 1998).

"It did not look like what we thought discrimination looked like."
(MIT, 1999)

This article looks briefly at the demography that provides crucial context for understanding why women's progress has been so glacially slow in academic positions in science and engineering. It then provides a brief introduction to the law and experimental social psychology relevant to understanding the "chilly climate."

---

*Paper presented at the National Academies Convocation on Maximizing the Success of Women in Science and Engineering: Biological, Social, and Organizational Components of Success, held December 9, 2005, in Washington, DC.

The ideal-worker schedule

The single most important statistic for understanding the dearth of academic women in science and engineering is that 95% of mothers aged 25 to 44 work less than 50 hours per week year-round[1] (Williams, 2000, p. 2). All a university has to do to drive most mothers off the tenure track is to define full time as 50-60 hours a week. Given that roughly 82% of women have children, driving away most mothers means driving away a very high percentage of women (Williams & Cooper, 2004).

Thus schedule alone goes a long way towards explaining why there are so few women in academic science and engineering (the "STEM disciplines"): the average workweek for scientists in education is 50.6 hours/week (NSF, 2005).

Just as most men want a career that does not require them to sacrifice conventional family life, so do most women.[2] For men, an academic career in the STEM disciplines typically does not require this sacrifice. For example, 68% of female physicists, but only 17% of male physicists, are married to other scientists, making the women much less likely than the men to have partners who can take care of the home front and leave the scientists with few responsibilities apart from working very long hours (McNeil, 1999).

Given that grant eligibility often is defined in terms of a certain number of years from PhD and that grant schedules typically do not allow time off for maternity leave, the STEM disciplines tend to idealize the worker who takes no time off for children. One result is that only 50% of tenured women academics in the STEM disciplines (but 70% of their male counterparts) have children (Mason and Goulden, 2002).

Indeed, the single-mindedness—and geographic mobility—required of academics in the STEM disciplines often mean that women sacrifice not only children but also marriage. This happens for two reasons. First, males are more likely to have a spouse who will follow them (Bielby, 1992); second, while career success in men is often considered an aphrodisiac (think Donald Trump), career success in women is a turn-off to many men.[3] Thus tenured women in academia are twice as likely as men to remain unmarried (Mason and Goulden, 2002).

---

[1]The percentage is for mothers aged 25-44, the key pre-tenure years, and for women who work year-round, which is what is required of academic scientists.

[2]Note the focus on *conventional* family life: most people's aspiration to marriage and children is not meant to endorse the view that other ways of building a life are inferior, or that many people create vibrant and vital families and other forms of intimacy that do not track the standard-issue spouse and kids model. Yet the fact remains that hegemony has a profound ability to shape the aspirations and imaginations of most people. (*See generally*, Gramsci, 1971.)

[3]This economy of desire reflects the eroticization of power in men, along with the accompanying instinct that there is "something wrong" when a man is married to a more accomplished woman (MacKinnon, 1987).

When women "opt out" of academic careers in the STEM disciplines (Belkin, 2003), it is often because they are faced with different choices than their male counterparts. Men are opting *into* a field that offers them a highly respected and intellectually challenging career that they can enjoy along with marriage and children. Women are opting *out of* a field that offers them a far less certain career path (due to glass-ceiling bias, described below), along with a high probability that they will have to sacrifice children and/or marriage along the way. Given their different contexts, that men and women make different choices is not (as we nonscientists like to say) rocket science.

Beyond culture and climate metaphors: glass-ceiling and maternal-wall stereotyping

In physics, "we select for assertiveness and single-mindedness."
(Georgi, 2000)

As suggested above, the ideal worker is designed around men's bodies (they need no time off for childbirth) and men's life patterns (American women still do 65-80% of the childrearing) (Sayer, 2001). The result is gender stereotyping in professional norms and everyday interactions.

One of the problems with the "chilly climate" and "culture" metaphors is that they provide little guidance for employers seeking to achieve a proportional representation of women.[4] WorkLife Law[5] (which I founded and direct) has worked hard to describe patterns of stereotyping in a way that provides clear guidance on what universities can do to eliminate the patterns of bias that plague the lives of many women.

The first, and most familiar, pattern is "glass-ceiling" bias many women encounter simply because they are women. WorkLife Law has documented that working women often encounter a second major form of gender bias, which we term the "maternal wall."[6] This term refers to the fact that many women who do not experience glass-ceiling bias find themselves facing discrimination triggered by family responsibilities once they become mothers (Biernat, Crosby and Williams, 2004; Williams and Segal, 2003).

In an era when the number of gender discrimination suits is falling, maternal wall suits are rising sharply. WorkLife Law has identified over 600 cases involving family responsibilities discrimination, a 400% increase in the last decade

---

[4]The chilly climate literature actually does better than the "climate" literature in describing the specific patterns that create problems for women.

[5]WorkLife Law is housed at UC Hastings College of the Law; http://www.worklifelaw.org.

[6]The "maternal wall" metaphor was introduced in Deborah J. Swiss and Judith P. Walker, *Women and the Work/Family Dilemma* (1993).

(Still, 2005). Because maternal wall cases have a higher win-rate than do civil rights cases in general (50% versus 27%), the potential for liability is substantial (Still, 2005). Thirty-seven verdicts and settlements have topped $100,000, with one over $11 million (Calvert, 2005).

A new role is emerging for gender stereotyping in maternal-wall cases. The traditional way of proving discrimination under Title VII of the Civil Rights Act of 1964 is by comparing the experience of a woman plaintiff to the experience of a "comparator": a similarly situated man. Yet in *Back* v. *Hastings-on-Hudson*, 365 F.3d. 107 (2d Cir. 2004), the Second Circuit Federal Court of Appeals allowed a case to go to trial despite the fact that the plaintiff lacked a comparator: She could not point to a similarly situated male school psychologist, because school psychologists typically are women. Yet Elana Back could, and did, identify significant evidence of gender stereotyping of mothers, notably the view (expressed by her principal and the head of human resources) that mothers are not committed to their careers.

Back clarified how stereotyping evidence can be used in federal discrimination cases, building on the glass-ceiling case of *Price Waterhouse* v. *Hopkins*, 490 U.S. 228 (1989). *Price Waterhouse*, through expert testimony of prominent social psychologist Susan Fiske, established that stereotyping evidence could be used to help a plaintiff prove sex discrimination. *Price Waterhouse* involved "hostile prescriptive stereotyping." A highly successful woman candidate for partner was told that she needed to "walk more femininely, talk more femininely, wear make-up," and "go to charm school"; in effect, behaving in a traditionally feminine manner was treated as a job requirement.

A third case added another key piece of the puzzle. In *Lust* v. *Sealy Inc.*, 383 F.3d 580 (7th Cir. 2004), the jury awarded the plaintiff over $1 million (later reduced). The lower court decision was upheld in a Seventh Circuit opinion that turned, in part, on a form of stereotyping called "cognitive bias." Cognitive bias is the bias that stems from the way our minds work in processing information (Krieger, 1995; Blasi, 2002). In *Lust*, the supervisor engaged in a form of cognitive bias called "attribution bias." When a man told the supervisor that he was interested in a promotion, the supervisor assumed that he was ready, willing, and able to move his family, whereas when Ms. Lust told the supervisor that she was interested in a promotion, he assumed (without asking her) that she would not move to take the job.

Price Waterhouse, *Back*, and *Lust* seem to signal that courts are ready, willing, and able to begin accepting evidence of gender stereotyping in discrimination cases. This interpretation seems all the more convincing given that while the opinion in one of the landmark 2004 cases (*Back*) was written by a liberal judge (Judge Guido Calibresi, former dean of Yale Law School), the opinions in the two others were written by leading conservative judges (Judges Richard Posner and Frank Easterbrook). These developments suggest a movement to accept stereotyping evidence that crosses ideological boundaries in an era when the

SECTION 2: SELECTED WORKSHOP PAPERS 153

federal courts are becoming increasingly conservative. Courts' increasing willingness to rely on stereotyping evidence is particularly important in academic cases because it provides an alternative method of proving discrimination in an employment context where comparators are often hard to find (AAUW, pp. 20-21).

Making bias visible in academic workplaces

Currently, the stereotyping literature is largely limited to experimental studies or meta-analyses confined to a specific theoretical approach. To be useful in guiding everyday workplace interactions, this literature must be consolidated and described in a way that is both scientifically responsible and readily understandable. What follows is WorkLife Law's attempt to do so.

The glass ceiling

The glass ceiling is composed of two distinct patterns: one makes it more difficult for women to establish themselves as competent; the other penalizes women for being *too* competent. Each pattern is described very briefly below, followed by an example from academia (from sciences or engineering, if one could readily be found) as well as references to the experimental social psychology literature.[7]

Trying twice as hard to achieve half as much: Patterns that make it more difficult for women to be perceived as competent

1. Women are judged on their accomplishments; men on their potential (Williams, 2003, pp. 416-417; McCracken, 2000, p. 159).
   - He's a "nascent scholar...soon to blossom"; she is unqualified due to lack of publications (*Lam v. University of Hawaii*, 59 Fair Empl. Prac. Cas. (BNA) 113 (1991))
2. "She's too feminine."
   - One faculty member voting against a female faculty member's tenure commented, "[her] problem in attracting graduate students was that she was too 'feminine' in that she was too 'unassuming, unaggressive, unassertive and not highly motivated for vigorous interpersonal competition.'" (*Zahorick v. Cornell University*, 729 F.2d 85, 89-90 (2d Cir. 1984); West, 1994, p. 132) Note the covert reliance on the association of femininity with low competence.

---

[7]The facts from the cases are told from the viewpoint of the plaintiff; in some cases, the court accepted the plaintiff's version of the facts and the plaintiff won; in other cases, either the case is still pending, or the plaintiff lost. Keep in mind, too, that evidence of stereotyping typically is only one element in meeting the legal standard for illegal discrimination.

- An accomplished female chemistry professor was characterized as "nice" and "nurturing" but not tenure material; a similarly accomplished man might well be seen as a good colleague and a good mentor.[8] (*Weinstock* v. *Columbia University*, 224 F.3d 33 (2nd Cir. 2000); Williams, 2004, pp. 45-47)

3. "He's skilled; she's lucky." Social psychologists long ago noted the tendency to attribute a man's successes to skill, while a woman's successes tend to be attributed to luck. This is an instance of "attribution bias" (Williams, 2003, p. 416; Swim & Sana, 1996).

4. "Anger is unseemly in a woman." Another example of attribution bias: An angry woman is a witch or a bitch, while an angry man is excused on the grounds that he understandably would not tolerate being "jerked around."

5. Recall bias. "Recall bias" is when women's mistakes are remembered forever but men's are soon forgotten (Williams, 2003, p. 417; Heilman, 1995, p. 6).

6. Gender-biased rewards. Men are sometimes given greater rewards than women for the same accomplishment (Williams, 2003, p. 418; Brewer, 1996, p. 63).

7. Objective rules are no guarantee of objectivity. Rules that are apparently objective can be framed around men or masculinity in ways that systematically disadvantage women. In addition, studies have documented *leniency bias*, when objective rules are applied rigidly to women, but flexibly to men (Williams, 2003, p. 415; Brewer, 1996, p. 65).

8. Polarized evaluations. In some institutions, woman superstars thrive, but women who are merely excellent are given much lower evaluations than similarly situated men. This key question is "whether a female schlemiel can do as well as a male schlemiel" (Williams, 2003, p. 418; Krieger, 1995, p. 1193; Yoder, 1994).

9. Are women isolated and "out of the loop"? Many academic departments in the STEM disciplines are overwhelmingly male. Studies have shown that women experience the problems of tokenism until women comprise 18-20% of a given workplace or department, a statistic that suggests that most women in science and engineering are at risk of being isolated and out of the loop (Williams, 2004; Biernat, 1998, p. 304; Taylor, 1981, p. 84).

- Social isolation is one reason single women without children in the bench sciences consider leaving academia (Mason and Goulden, 2002).

10. Is the job defined in terms of masculine patterns? Recall the quote that academic physics is defined in terms of assertiveness and single-mindedness (Georgi, 2000). *Single-mindedness*, as noted above, is a polite way of describing the requirement that, to be successful, a scientist must either eschew family life or enjoy a flow of domestic services from a spouse that is common among men but

---

[8]The case cited contains only the characterizations of Weinstock, a woman professor of chemistry; the others are added to highlight how a similarly situated man might be described.

rare among women. *Assertiveness* signals that the personality characteristics assumed to be required for success in science often track those traditionally associated with men. When a job is defined in terms that confuse masculinity with job qualifications, bias against women occupants of the job is likely to be commonplace (Williams, 2003, pp. 408-409; Heilman, 1993, p. 280).

11. Mother, princess, pet. An extremely important point, rarely understood, is that environments plagued by gender bias do not affect all women in the same way. Women are stereotyped by subtype, not as a generalized group (Glick and Fiske, 2001, p. 113; Williams, 2003, p. 419). In some departments, women who play stereotypically feminine roles are taken into the in-group and do well, while women who do not adhere to traditionally feminine behavior are stigmatized. Established feminine roles documented by social psychologists include the mother, who may take charge of departmental teas and comfort; the princess, who aligns herself with powerful men; the pet, who is nonthreatening and cuddly; and Ms. Efficiency, who acts as a glorified but subservient secretary (Taylor, 1981, p. 84).

- "[Faculty] don't realize that often they—men and women—expect women to make [their colleagues] feel comfortable, and [they] don't expect men to make [their colleagues] feel comfortable." (AAUW, 2004, p. 35)
- "[My wife did not get tenure because] she had not played at being a good daughter to the older and more traditional men on the faculty, giggling at their jokes and massaging their egos." (West, 1994, p. 145, quoting Robert Reich)

12. Subtype revisited: feminist. Stereotype content studies show negative associations with the label "feminist". (Glick and Fiske, 2002) When a woman stands up for herself or women's rights and is characterized as "shrill," "a feminazi," or a "fanatical feminist," this is evidence of gender bias. (Note that we no longer hear claims that African-Americans "have it coming" if they are "too uppity": Women, too, should be able to stand up for their group without being demonized.)

Patterns that penalize women for being too competent

Sometimes women are disadvantaged because they do not conform to their colleagues' image of how women should behave.

1. "He's assertive, she's aggressive" (Taylor, 1981, p. 103).
- According to Martin Snyder of the American Association of University Professors, recent collegiality cases "all came down to the same thing. They're all-male dominated departments that hadn't tenured a woman in a long time, or ever, and there's some language about how the woman 'just doesn't fit in.' What comes through is that these are aggressive women who are seen as uppity." (Lewin, 2002)
- "To get ahead here [at MIT], you have to be so aggressive. But if women are too aggressive they're ostracized . . ." (Haak, 1999)

2. Catch-22. "... and if they're not aggressive enough they have to do twice the work." (Haak, 1999). The leading case, *Price Waterhouse* v. *Hopkins*, 490 U.S. 228, 250 (1989), christened this a "Catch-22" and treated it as potentially illegal gender discrimination.

3. Ambivalent sexism. In some departments, a woman may have to choose between being liked, but not respected, or respected, but not liked (Glick and Fiske, 1999). Neither path leads to tenure.

- Her colleagues were "indifferent, if not hostile, to her accomplishments. . . . Ironically, her prestige and status outside the university grew as her status at the university diminished and became more precarious." (AAUW, 2004, p. 58)

4. She's a "bitter, selfish" effective manager. Women managers tend to be categorized as either unqualified because they are ineffective managers or as unqualified because they are effective but have personality problems (Heilman, 2001). This again stems from the fact that the qualities associated with traditionally masculine professions such as scientist, engineer, and manager closely track the qualities associated with a typical man, but not those associated with a typical woman.

- "There might be a perception that, as a woman, [a woman candidate] should have a warm and fuzzy personality. [Name of candidate] is not a warm and fuzzy person...". (AAUW, 2004, p. 35)

5. "She's a shameless self-promoter; he knows his own worth". Particularly if a woman colleague is socially isolated, she may well have no mentor who can highlight her accomplishments. Yet if she does so herself, studies show that self-promotion in women may well trigger negative reactions not triggered by self-promotion in men (Eagley and Karau, 2002, p. 584; Williams, 2003, p. 425).

6. Sexual harassment of successful women. Sexual harassment is one way sexist men have of controlling women they find threatening. This is typically a "no win" situation for women: A survey by the American Management Association found that even if the woman is the victim, she is just as likely as the offender to be dismissed or transferred (Grimsley, 1996).

The maternal wall

1. Jobs defined in masculine terms, revisited. As noted above, the ideal academic worker reflects a template designed around masculinity (Williams, 2000). When the ideal worker designed around men, no wonder so few women measure up.

2. An unsuitable job for a mother: role incongruity. Particularly when the ideal worker is defined as requiring a 24/7 commitment, being an academic may be seen as inconsistent with being a mother. (Williams, 2003, pp. 430-431; Etaugh and Gilomen, 1989; Eagly and Steffen, 1986, p. 254; Kobrynowicz and Biernat, 1997, p. 593).

- In a case with a reported settlement of nearly $500,000, a memo was circulated saying, "As a mother of two infants, she had responsibilities that were incompatible with those of a full-time academician."[9] (Schneider, 2000).

3. "Loose lips": prescriptive stereotyping, benevolent or hostile (Glick and Fiske, 2001; Burgess and Borgida, 1999). Hostile prescriptive stereotyping prescribes that mothers should stay home with their children. Kinder, gentler "benevolent" prescriptive stereotyping is when a department attempts to do a mother a favor by, for example, not allocating travel funds because they assume that children need their mothers and that mothers will not want to travel. (Note that the solution is to simply ask the mother, rather than making assumptions.)

- A department chair argued that a woman did not need her job as much as a man because she was married (and presumably her husband could support her) (AAUW, 2004, p. 5).

4. Maternal wall attribution bias. An absent man is giving a paper; an absent woman is assumed to be home with her children (even if she is at a conference). (Williams, 2003, pp. 433-434; Eagley and Karau, 2002, p. 589; Kennelly, 1999, p. 176)

5. Maternal wall leniency bias. Mothers are held to longer hours and to higher performance and punctuality standards. (Williams, 2003, p. 433; Correll and Benard, 2005)

6. Negative competence assumptions associated with motherhood. A 2005 study found that "relative to other kinds of applicants, mothers were rated as less competent, less committed, less suitable for hire, promotion, and management training, and deserving of lower salaries." (Correll & Benard, 2005)

- According to one scientist, "The perception of me [after] having a child is that my profession is not the priority anymore. . . ." (Bombardieri, 2005)
- "Several female professors believed that pregnancy had hampered their chances for tenure because they were viewed as less serious about or committed to their careers" (AAUW, 2004, p. 27).

Earlier stereotype content studies show that, although businesswomen are rated as high in competence, similar to businessmen, housewives are rated as extremely low in competence, alongside stigmatized groups such as the elderly, blind, "retarded," and disabled (Glick and Fiske, 2002; Eckes, 2002, p. 110).

- "If you…have your child on campus, colleagues who recognize you when you are by yourself now only see you as a walking uterus and ignore you." (Mason, 2003, p. 2)

7. Part-time work is an independent trigger for negative competence assumptions. Women who work part-time may get the worst of both worlds: They are seen as less competent workers than women who work full time and less compe-

---

[9]It was unclear whether the memo referred to the tenure candidate or the department chair, who was also a woman.

tent mothers than homemakers[10] (Williams, 2004, p. 388). (Note that women who trigger negative competence assumptions often are held to a higher standard, which is illegal under federal antidiscrimination law, as discussed below.)

- A female professor claimed that her pregnancy, which has caused her to switch to part-time tenure track (among other things) led her institution to hold her to a higher standard than similarly situated males (AAUW, 2004, p. 50).

8. The maternal wall can trigger gender wars among women. Often the maternal wall triggers fights among professional women, which can be particularly acute in the sciences given that 50% of women academic scientists have no children (Mason and Goulden, 2002). Some of these women are child*free*—cultural entrepreneurs who are trying to invent an image of a full, female life without children, and may feel that mothers who demand "special treatment" are reinforcing stereotypes that women can't measure up (Burkett, 2000). Others are child*less*, regretful they did not have children. They may well ask why mothers should "have it all" when they themselves had to make a choice between career and having a family (ignoring the fact that most male academics "have it all" as a matter of course) (Hewlett, 2002). When the maternal wall pits women against women, this is a *result* of gender discrimination; yet it is often cited as evidence that "this is not a gender problem."

9. Family responsibilities discrimination against men. The technical name for maternal wall discrimination is family responsibilities discrimination (FRD), because it can affect men as well as women. When men seek to take on traditionally feminine caregiving roles, they may well suffer even more severe consequences than do women. A study of over 500 employees found that, when compared to mothers, fathers who took a parental leave were recommended for fewer rewards and viewed as less committed, and fathers with even a short work absence due to a family conflict were recommended for fewer rewards and had lower performance ratings (Dickson, 2004).

- An untenured professor told his mentor that he did not dare even to ask about parental leave, much less take it (Source: confidential).
- A father was denied a child-rearing leave routinely available to women (*Shafer* v. *Board of Public Education*, 903 F.2d 243, 244 (3rd Cir. 1990)).

Chilly climate patterns may be evidence of illegal gender discrimination

While this is not the forum for a full discussion of the potential for legal liability, a brief review of the applicable law serves to highlight that many pat-

---

[10]Eagly and Stephen's study appears to be contradicted by another study that reports that women who switch from full-time to part-time schedules are not viewed as lower in competence than women in full-time work. (Etaugh and Moss, 2001).

terns that comprise the "chilly climate" are potentially illegal. Where available, an academic example or two is provided.

Title VII of the Civil Rights Act of 1964: Disparate treatment
(42 USC § 2000e-2 (a)(1))

As noted above, the first way of proving disparate treatment is to show that a woman plaintiff is treated differently than a comparator, as when a man is hired based on his potential, but a woman with the same qualifications is not hired because she is judged strictly on what she has already accomplished. As also noted above, even in the absence of a comparator (or in addition to comparator evidence), a plaintiff can rely on evidence of gender stereotyping (*Back*, 2004).

- A psychology professor called the WorkLife Law hotline. She had outstanding job evaluations from peers and students—until she had a baby. After her baby was born, she still got high evaluations from students, but not from her colleagues. The head of her department engaged in intense scrutiny of her office hours, although he did so for no other member of the department, a classic example of a woman being treated differently than the relevant male comparators. In addition, the department head and members of her tenure committee stated that, "[P]eople who prioritize family do not make tenure," a statement that reflects maternal wall stereotyping.

Title VII of the Civil Rights Act of 1964: Disparate impact
(42 USC § 2000e-2 (a)(2))

A facially neutral policy that disproportionately impacts women may be illegal if it is not justified by business necessity. Even if the policy *is* justified by business necessity, the employer may be liable if the plaintiff can show that an alternative less discriminatory policy could accomplish the same goal. (42 USC § 2000e-2 (k)(1)(A). An example of a policy that has a disparate impact on women is a university policy that denies proportional benefits to professors on part-time tenure track; the university would argue business necessity.

Title VII of the Civil Rights Act of 1964: Sexual harassment
(42 USC § 2000e-2)
Quid pro quo (your body or your job)

- A senior professor remarked to a woman colleague that her refusal of his sexual advances was "no way to get tenure." (AAUW, 2004, p. 61)

Hostile environment

- Some women academics have reported that they themselves are sexually harassed when they blow the whistle against other faculty members who have harassed women. (AAUW, 2004, p. 13)
- One woman professor alleged hostile work environment when a department celebrated a colleague's birthday with a "boob cake." A male faculty called her a lesbian because she turned down dates with male faculty members. Women signed up for the department athletic team only to be turned down and insulted. The plaintiff was given unusually heavy teaching and service loads but was not allowed to teach graduate courses. Women were asked to appear at many functions to present an image that the department had a substantial number of women on the faculty (which they did not) (AAUW, 2004, p. 29).

Title VII of the Civil Rights Act of 1964: Constructive discharge

Constructive discharge occurs when an employer imposes intolerable working conditions stemming from unlawful discrimination or harassment that would compel a reasonable person to quit (Center for WorkLife Law, 2006, pp. 29-31).

Title VII of the Civil Rights Act of 1964: Retaliation (42 USC § 2000e-3(a))

An employer is prohibited from retaliating against women for engaging in conduct Title VII protects. In some jurisdictions, retaliation is defined narrowly, and covers only situations in which a woman reports alleged harassment and discrimination to a civil rights enforcement agency. In others, retaliation is defined more broadly, and also covers reporting discrimination or harassment to management, having a lawyer write a letter alleging discrimination, refusing to settle a prior claim of discrimination, or stating an intention to file a civil rights complaint (Center for WorkLife Law, 2006, pp. 33-37). Retaliation may well be easier to prove than the underlying discrimination case.

- A woman professor was denied tenure after she advocated better treatment of female graduate students and faculty; in particular, she publicly objected to the disproportionate service responsibilities assigned to women (AAUW, 2004, p. 17).
- A women professor was retaliated against when she complained of pay inequity in violation of the EPA and of her denial of tenure (AAUW, 2004, p. 21).
- A professor who objected to the treatment of women after childbirth was branded a troublemaker and fired at the first opportunity (AAUW, 2004, p. 26).

Equal protection (a constitutional claim) (42 USC. § 1983)

Academics who teach in public universities can recover if they can prove that men were disadvantaged as compared with women, as when leave is routinely offered to women but men are forbidden or severely discouraged from taking it. Women in public universities also can sue if they are not given equal protection of the law.

Equal Pay Act

It is illegal to pay higher salaries to men than to women doing "equal work" in jobs that require substantially "equal skill, effort, and responsibilities . . . under equal working conditions" (29 USC § 206(d)(1)). One federal case, *Lovell* v. *BBNT Solutions, LLC,* 295 F. Supp. 2d 611 (E.D. VA. 2003), refused to apply a categorical rule excluding a part-time chemist from being compared to full-time chemists, in a ruling that suggests that professors on part-time tenure track should be paid the proportion of their salary equal to the proportion of a full-time schedule they work (for example, 75% pay for a 75% workload).

- When a female professor was hired she was told that the institution was prohibited from paying her more than a specified base salary and $5,000 as an administrative stipend, only to later discover that other professors were paid more (AAUW, 2004, p. 22).

Pregnancy Discrimination Act (PDA) (42 USC § 2000e-(k))

Employers are required to treat pregnant professors "the same" as other workers whose ability to work is similar. Evidence of a violation of the PDA includes stereotyping pregnant women as incompetent or not committed to their careers, stripping a pregnant woman of duties and opportunities, or imposing conditions on her that are not applied to nonpregnant employees.

- One professor alleged that, in her department, pregnant professors' chances for tenure were hampered because they were viewed as less serious, less committed to their careers, and because of animosity stemming from the way their teaching responsibilities were reallocated to their colleagues (AAUW, 2004, p. 27).
- Another professor alleged that pregnant women were not treated the same as other professors because, while the university gave assistant professors an extra year for a variety of reasons, it refused to stop the tenure clock after she had a baby, thereby making her record appear weaker than those of her colleagues. Another colleague had the clock stopped for a year but was not told that the university expected an extra year's publications despite their agreement to stop the clock.
- An institution required pregnant women to choose between the parental

leave benefit available to men and women (a course release) and their 6- to 8-week maternity disability leave; other faculty members were granted disability leave without being required to sacrifice another benefit in order to obtain it.

Family Medical Leave Act of 1993 (FMLA) (29 U.S.C. § 2601)

Denial of leave: The FMLA gives professors the legal right to up to 12 weeks of unpaid leave per year if the employee or her child, partner, or parent has a serious health condition, or if she has or adopts a child. Giving leave is *mandatory*.

Interference with leave: In addition to denying leave, covered employers are prohibited from interfering with leave.
- Female faculty hesitated to take the four-week disability leave immediately after birth, sensing "pressure . . . not to take it." Candidates for leave were told that "[t]aking a four week maternity leave may be seen by some members of the committee as a lack of commitment to career, and a premeditated plan to [impose on their colleagues]." In this context, a female faculty member was in the classroom five days after she gave birth[11] (AAUW, 2004, p. 28).

Americans with Disabilities Act (ADA) (42 USC § 12101)

Employees may not be discriminated against because they are caring for a family member whose illness or disability is covered by the ADA.

Title IX (20 USC § 1681)

The federal government can block all federal funding to an academic institution if it discriminates on the basis of sex, parental status, primary caregiver status, or pregnancy in its educational programs. This "atom bomb" sanction is rarely used. Title IX also allows professors to sue for discrimination, although some jurisdictions limit their ability to receive damages (Center for WorkLife Law, 2006, pp. 63-64).

Conclusion

This article introduces a very different language for talking about gender bias in the STEM disciplines than the traditional metaphors of the "chilly climate" or the "academic culture" (Sandler, et al., 1996; Stanford, 1993; Mervis, 2002; Trower and Chait, 2002). While these metaphors have played a useful role, it is

---

[11]The actual quote ended, "to cause an imposition."

time to re-examine them. Developments in experimental social psychology and the law offer more concrete and effective guidance on how to improve the status of women.

The approach outlined here, I believe, is important for two reasons. First, recent studies show that, while diversity training shows no demonstrable effect in increasing the numbers of minorities and women, litigation—or the potential for it—*does* often spur institutional change (Kalev, Dobbin, and Kelly, 2005; Pedriana and Stryker, 1997). Second, the booming literature on cognitive bias shows that while stereotype *activation* is automatic, stereotype *application* can be controlled (Sommers and Ellsworth, 2001; Blair, 2002). Before stereotypes can be controlled, however, they must first be recognized. The approach outlined in this article has the potential to spur that process by making stereotyping and bias visible in the STEM disciplines.

References

American Association of University Women (2004). *Tenure denied: cases of sex discrimination in academia.* Washington, DC: AAUW Education Foundation and Legal Advocacy Fund, http://www.case.edu/president/aaction/TenureDenied.pdf.

Back v. Hastings-on-Hudson Union Free School District, 365 F.3d. 107 (2d Cir. 2004).

L Belkin (2003). The opt-out revolution. *New York Times Magazine* October 26:42.

WT Bielby and D Bielby (1992). I will follow him: family ties, gender-role beliefs, and reluctance to relocate for a better job. *American Journal of Sociology* 97(5):1241-1267.

M Biernat, CS Crandall, LV Young, D Kobrynowicz, and SM Halpin (1998). All you can be: stereotyping of self and others in a military context. *Journal of Personality and Social Psychology* 75(2):301-317.

M Biernat, FJ Crosby, and JC Williams (2004). The maternal wall. *Journal of Social Issues* 60(4):675-683.

IV Blair (2002). The malleability of automatic stereotypes and prejudice. *Personality and Social Psychology Review* 6:242-261.

G Blasi (2002). Advocacy against the stereotype: Lessons from cognitive social psychology. *University of California, Los Angeles Law Review* 49(5):1241-1281.

M Bombardieri (2005). Reduced loan lets faculty meld family, tenure track. *Boston Globe*. October 4.

MB Brewer (1996). In-group favoritism: the subtle side of intergroup discrimination. In *Codes of conduct: behavioral research into business ethics,* eds. DM Messick and AE Tenbrunsel, New York: Russell Sage Foundation.

E Burkett (2000). *The baby boom.* New York: The Free Press.

D Burgess and E Borgida (1999). Who woman are, who women should be: descriptive and prescriptive gender stereotypes in sex discrimination. *Psychology, Public Policy and Law* 5:665-687.

C Calvert (2005). E-mail to Joan C. Williams. November 7, 2005.

Center for WorkLife Law (2006). WorkLife Law's guide to caregiver discrimination (forthcoming).

Civil Rights Act of 1964, Title VII. 42 USCS § 2000e-2.

SJ Correll and S Benard (2005). *Getting a job: Is there a motherhood penalty*? Presentation at American Sociological Association Annual Meeting, August 15, 2005, Philadelphia, PA. http://sociology.princeton.edu/programs/workshops/Correll_Benard_manuscript.pdf.

CE Dickson (2004). The impact of family supportive policies and practices on perceived family discrimination, (dissertation).

AH Eagly and SJ Karau (2002). Role incongruity theory of prejudice toward female leaders. *Psychology Review* 109:537-598.

AH Eagly and VJ Steffen (1986). Gender stereotypes, occupational roles, and beliefs about part-time employees. *Psychology of Women* 10:252-262.

T Eckes (2002). Paternalistic and envious gender prejudice: testing predictions from the stereotype content model. *Sex Roles* 47:99-114.

Equal Pay Act. 29 USC § 206 (d)(1).

Equal Protection. Section 1983, 42 USC § 1983.

C Etaugh and G Gilomen (1989). Perceptions of mothers: effects of employment status, marital status, and age of child. *Sex Roles* 20:59-70.

C Etaugh and C Moss (2001). Attitudes of employed women toward parents who choose full-time or part-time employment following their child's birth. *Sex Roles* 44:611-619.

Family and Medical Leave Act of 1993. (29 U.S.C. § 2601).

H Georgi (2000). Is their an unconscious discrimination against women in science? *The American Physical Society News* 9(1):27-30. http://www.aps.org/apsnews/0100/010016.cfm.

P Glick and ST Fiske (2002). A model of (often mixed) stereotype content: competence and warmth respectively follow from perceived status and competition. *Journal of Personality and Social Psychology* 82:878-902.

P Glick and ST Fiske (2001). An ambivalent alliance: hostile and benevolent sexism as complementary justifications for gender inequality. *American Psychology* 58:109-118.

P Glick and ST Fiske (1999). (Dis)respecting versus (dis)liking: Status and interdependence predict ambivalent stereotypes of competence and warmth. *Journal of Social Issues* 55:473-489.

Griggs v. *Duke Power Co.*, 401 U.S. 424 (1971).

American Management Association survey cited in KD Grimsley (1996). Co-Workers Cited in Most Sexual Harassment Cases; Management Group's Study Disputes Stereotype. *The Washington Post*, June 14, p. D01.

L Haak (1999). Women in Neuroscience: The first twenty years. *Journal of the History of Neuroscience* 11:70-79.

S Hewlett (2002). *Creating a life: professional women and the quest for children*. New York: Talk Miramax Books.

ME Heilman (2001). Description and prescription: How gender stereotypes prevent women's ascent up the organizational ladder. *Journal of Social Issues* 57:657-674.

ME Heilman (1995). Sex stereotypes and their effects in the workplace: What we know and what we don't know. *Journal of Social Behavior and Personality* 10 (3):3-26.

ME Heilman (1993). Sex bias in work settings: The lack of fit model. In *Research in Organizational Behavior*, eds. LL Cumings and BM Staw. Greenwich, CT: JAI Press.

A Kalev, F Dobbin, and E Kelly (2005). *Best Practices or Best Guesses? Diversity Management and the Remediation of Inequality*. Working Paper. Department of Sociology, Harvard University.

I Kennelly (1999). That single mother element: How white employers typify black women. *Gender and Society* 14:168-192.

D Kobrynowicz and M Biernat (1997). Decoding subjective evaluations: How stereotypes provide shifting standards. *Journal of Experimental Social Psychology* 33:579-601.

LH Krieger (1995). The content of our categories: A cognitive bias approach to discrimination and equal employment opportunity. *Stanford Law Review* 47:1161-1248.

Lam v. *University of Hawaii*, 59 Fair Empl. Prac. Cas. (BNA) 113 (1991).

T Lewin (2002). "Collegiality" as a tenure battleground. *New York Times*, July 12, p. A12.

Lovell v. *BBNT Solutions*, LLC, 295 F. Supp. 2d 611 (E.D. Va. 2003).

Lust v. *Sealy*, 383 F.3d 580 (7th Cir. 2004).

C MacKinnon (1987). *Feminism unmodified: discourses of life and law*. Cambridge, MA: Harvard University Press.

MA Mason (2003). UC Berkeley faculty work and family survey: preliminary findings, http://universitywomen.stanford.edu/reports/UCBfacultyworknfamilysurvey.pdf.

MA Mason and M Goulden (2002). Do babies matter? The effect of family formation on the lifelong careers of academic men, and women. *Academe*, November-December. http://www.aaup.org/publications/Academe/2002/02nd/02ndmas.htm.

Massachusetts Institute of Technology (1999). A study on the status of women faculty in science at MIT. *The MIT Faculty Newsletter* 11 (4):14-26, http://web.mit.edu/fnl/women/women.html.

DM McCracken (2000). Winning the talent war for women: Sometimes it takes a revolution. *Harvard Business Review*, http://hbswk.hbs.edu/item.jhtml?id=1840&t=organizations.

L McNeil (1999). Dual-career-couples: Survey results. http://www.physics.wm.edu/~sher/survey3.html.

J Mervis (2002). Can equality in the sports be repeated in the lab? *Science* 298:356.

National Science Foundation (2005). All in a week's work: average work weeks of doctoral scientists and engineers, http://www.nsf.gov/statistics/infbrief/nsf06302/nsf06302.pdf.

N Pedriana and R Stryker (1997). Political culture wars 1960s style: Equal employment opportunity-affirmative action law and the Philadelphia plan. *American Journal of Sociology* 103(3):633-691.

Pregnancy Discrimination Act 42 USC § 2000e (k).

Price Waterhouse v. *Hopkins*, 490 U.S. 228 (1989).

BR Sandler, LA Silverberg, and RM Hall (1996). *The Chilly Classroom Climate: A Guide to Improve the Education of Women*. Washington, DC: The National Association of Women in Education.

LC Sayer (2001). *Time, use, gender, and equality: Differences in men's and women's market, non-market, and leisure time*. Unpublished doctoral dissertation, University of Maryland, College Park.

A Schneider (2000). University of Oregon settles tenure lawsuit over maternity leave. *The Chronicle of Higher Education* (July 21), p. A12.

Shafer v. *Board of Public Education*, 903 F.2d 243 (3rd Cir. 1990).

SR Sommers and PC Ellsworth (2001). White juror bias: An investigation of racial prejudice against Black defendants in the American courtroom. *Psychology, Public Policy, and Law* 7:201-229.

Stanford University (1993). Report of the Committee on the Recruitment and Retention of Women Faculty at Stanford. M. Strober, Chair.

M Still (2005). E-mail to Joan C. Williams, December 7.

AV Sullivan (2005). Breaking anonymity: The chilly climate for women faculty by the chilly collective, http://fcis.oise.utoronto.ca/~hep/Sullivan.html.

JK Swim and LJ Sana (1996). He's skilled, she's lucky: a meta-analysis of observers' attribution for women's and men's successes and failures. *Personality and Social Psychology Bulletin* 22(5):507-519.

DJ Swiss and JP Walker (1993). *Women and the Work/Family Dilemma*. New Jersey: John Wiley and Sons.

SE Taylor (1981). A categorization approach to stereotyping. In *Cognitive Processes in Stereotyping and Intergroup Behavior*, ed. DL Hamilton. New Jersey: Lawrence Erlbaum Associates.

CA Trower and RP Chait (2002). Faculty diversity: Too little for too long. *Harvard Magazine* (March-April):33-38. http://www.harvardmagazine.com/on-line/030218.html.

V Valian (1998). *Why so slow? The advancement of women*. Cambridge, MA: MIT Press.

Weinstock v. *Columbia University*, 224 F.3d 33 (2nd Cir. 2000)

M West (1994). Gender bias in academic robes: The law's failure to protect women faculty. *Temple Law Review* 67:68-178.

JC Williams (2004). Hitting the maternal wall. *Academe*, http://www.aaup.org/publications/Academe/2004/04nd/04ndwill.htm.

JC Williams and HC Cooper (2004). The public policy of motherhood. *Journal of Social Issues* 60(4):849-865.

JC Williams and N Segal (2003). Beyond the maternal wall: relief for family caregivers who are discriminated against on the job. *Harvard Women's Law Journal* 26:77-162.

JC Williams (2003). The social psychology of stereotyping: Using social science to litigate gender discrimination cases and defang the "cluelessness" defense. *Employee Rights and Employment Policy Journal* 7(2):401-458.

JC Williams (2000). *Unbending gender: Why work and family conflict and what to do about it.* New York, NY: Oxford University Press.

JD Yoder (1994). Looking beyond numbers: the effect of gender status, job prestige, and occupational gender-typing on tokenism processes. *Social Psychology Quarterly* 57:150-159.

Zahorik v. *Cornell University*, 729 F.2d 85, 89-90 (2d Cir. 1984)

SOCIAL INFLUENCES ON SCIENCE AND ENGINEERING CAREER DECISIONS[1]

Yu Xie
University of Michigan

Abstract

Our study on the career processes and outcomes of women in science has four major components. First, rather than focusing on specific segments of a science/engineering (S/E) career, we studied the entirety of a career trajectory. Second, we analyzed seventeen large, nationally representative datasets. Third, we tried to be as objective and "value-free" as possible and to emphasize empirical evidence. Finally, we based the book on a life-course approach, a combination of special methodological perspectives which recognize the following phenomena:

a. Interactive effects across multiple levels, *such as the individual level, the family level, and the school level. Individuals do not live or work in isolation from one another.*

b. Interactive effects across multiple domains, *such as education, family, and work. What we do in one domain of our lives affects what we do in other domains.*

c. Individual-level variations in career tracks *resulting from differences among individuals, even those with the same demographic characteristics.*

d. The cumulative nature of the life course. *What happened before affects what happens now, and what is happening now affects what comes next. This is also called "path-dependency." Because of path dependency, small differences at particular points in time can deflect trajectories and subsequently generate large differences in career outcomes.*

---

[1]This presentation is based on the book Yu Xie co-authored with Kimberlee Shauman entitled, *Women in Science: Career Processes and Outcomes,* published by Harvard University Press in 2003.

*Paper presented at the National Academies Convocation on Maximizing the Success of Women in Science and Engineering: Biological, Social, and Organizational Components of Success, held December 9, 2005, in Washington, DC.

The life course approach places a high demand on data. Ideally, we would like to have longitudinal data over the entire career of many scientists and non-scientists. We looked very hard but were not able to find a perfect data set. Lacking such a data set, we were still able to carry out our study by piecing together many datasets to paint a composite picture of gender differences in science careers, a method which is called *synthetic cohort* in demography. Figure 2-12 shows the data sets that we used to look at different segments of science/engineering careers.

Needless to say, our study contains complicated and nuanced analyses. These analyses led us to conclude that women's severe underrepresentation in science and engineering is an extremely complex social phenomenon that defies any attempt at simplistic explanations. Due to the complex and multi-faceted nature of women scientists' career processes and outcomes, especially how these processes and outcomes affect, and are affected by, other life course events such as marriage and childbearing, we were uncomfortable recommending concrete policy interventions intended to increase women's representation in science and engineering No single explanation or hypothesis testing should or could substitute for the richness of the empirical results from these analyses, though we did consider and reject several widely accepted hypotheses, as the following discussion shows.

The "Critical Filter" Hypothesis

One longstanding hypothesis in the literature is that women are less likely to pursue science/engineering careers because they are handicapped by deficits in high school mathematics training. In a classic statement of this position, Sells (1980) claims that "[a] student's level of high-school mathematics achievement acts as a critical filter for undergraduate college admission for blacks and limits choices of an undergraduate major for women in general once they are admitted to college." This hypothesis is appealing for its simplicity and the clear remedy it implies.

From our research, we find that the gender gap in average mathematics achievement is small and has been declining since the 1960s (see Table 2-5). The numbers in Table 2-1 are mean gender differences in math achievement scores (in standard deviation units). The declining trend shown in Table 2-5 casts doubt on the interpretation that the gender gap in math achievement reflects innate, perhaps biological, differences between the sexes. We also find that the gender gap in representation among top achievers remains significant (see Table 2-6). This finding was cited by Harvard President Larry Summers in his remarks at an NBER conference on January 14, 2005, which made international news. However, President Summers failed to cite the following finding: gender differences in neither average nor high achievement in mathematics explain gender differences in the likelihood of majoring in science/engineering fields.[2]

---

[2] See Xie and Shauman (2003). Ibid, Chapters 3 and 4.

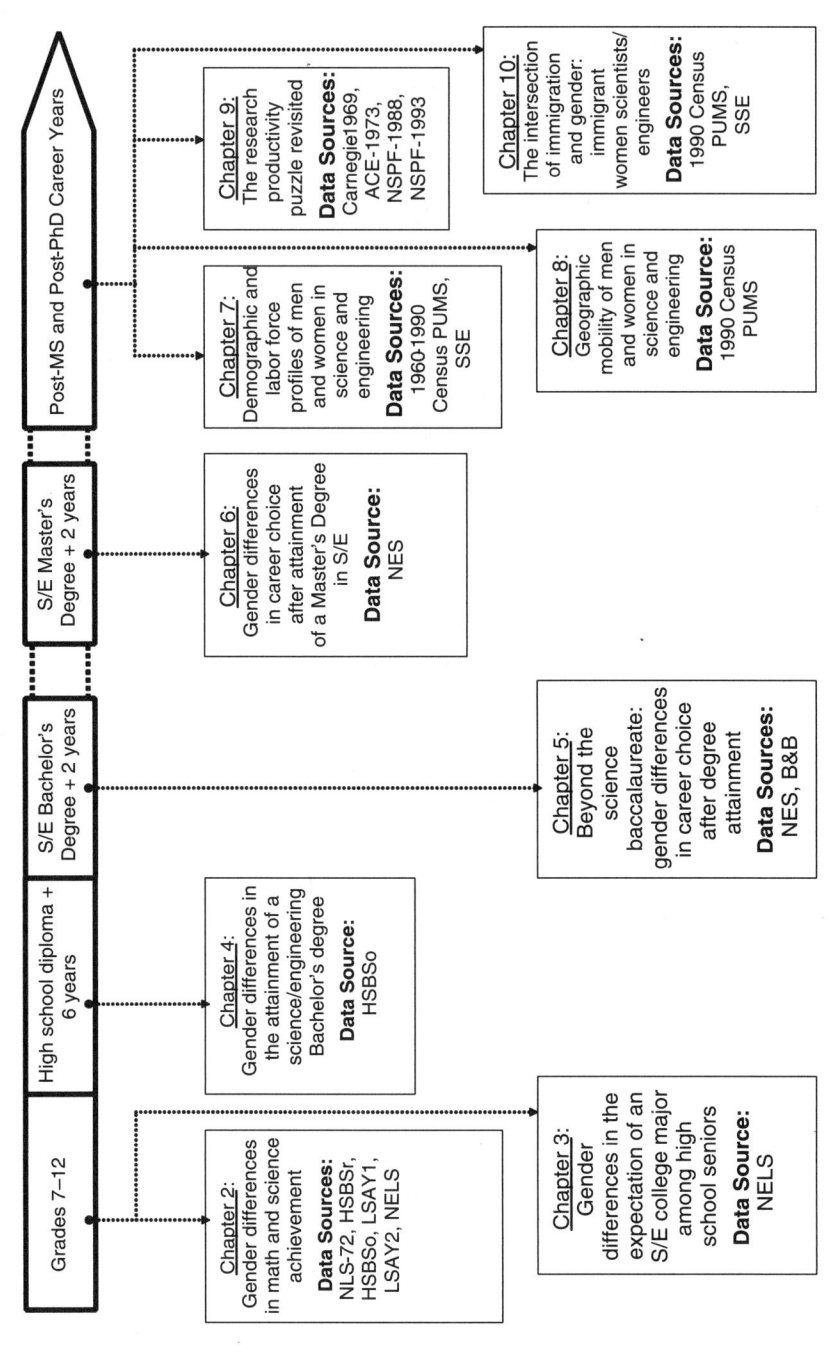

FIGURE 2-12 Synthetic cohort life course, career processes, and outcomes examined, and data sources.

TABLE 2-5 Standardized Mean Gender Difference of Math Achievement Scores Among High School Seniors by Cohort

School Cohort	Mean Difference (d)	Data Source
1960	−0.25***	NLS-72
1968	−0.22***	HSBSr
1970	−0.15***	HSBSo
1978	−0.13**	LSAY1
1980	−0.09***	NELS

*p<.05 **p<.01 ***p<.001 (two-tailed test), for the hypothesis that there is no mean difference between males and females.

TABLE 2-6 Female-to-Male Ratio of the Odds of Achieving in the Top 5% of the Distribution of Math Achievement Test Scores Among High School Seniors by Cohort

School Cohort	Achievement Ratio	Data Source
1960	0.45***	NLS-72
1968	0.47***	HSBSr
1970	0.48***	HSBSo
1978	0.25***	LSAY1
1980	0.60***	NELS

*p<.05 **p<.01 ***p<.001 (two-tailed test), for the hypothesis that there is no mean difference between males and females.

The Pipeline Paradigm

A dominant perspective in the literature on women in science is the "pipeline" paradigm. According to this paradigm, the process of becoming a scientist can be conceptualized as a pipeline, called the "science pipeline," which is essentially a developmental process. Change in the developmental process along the life course is unidirectional—leaving science versus staying in science.

However, we find career processes to be fluid and dynamic. Exit, entry, and reentry are real possibilities. Many persons, especially women, become scientists through complicated processes rather than by just staying in the pipeline. Also, we show that participation gaps are greatest at the transition from high school to college. This is illustrated in Figure 2-13.

170

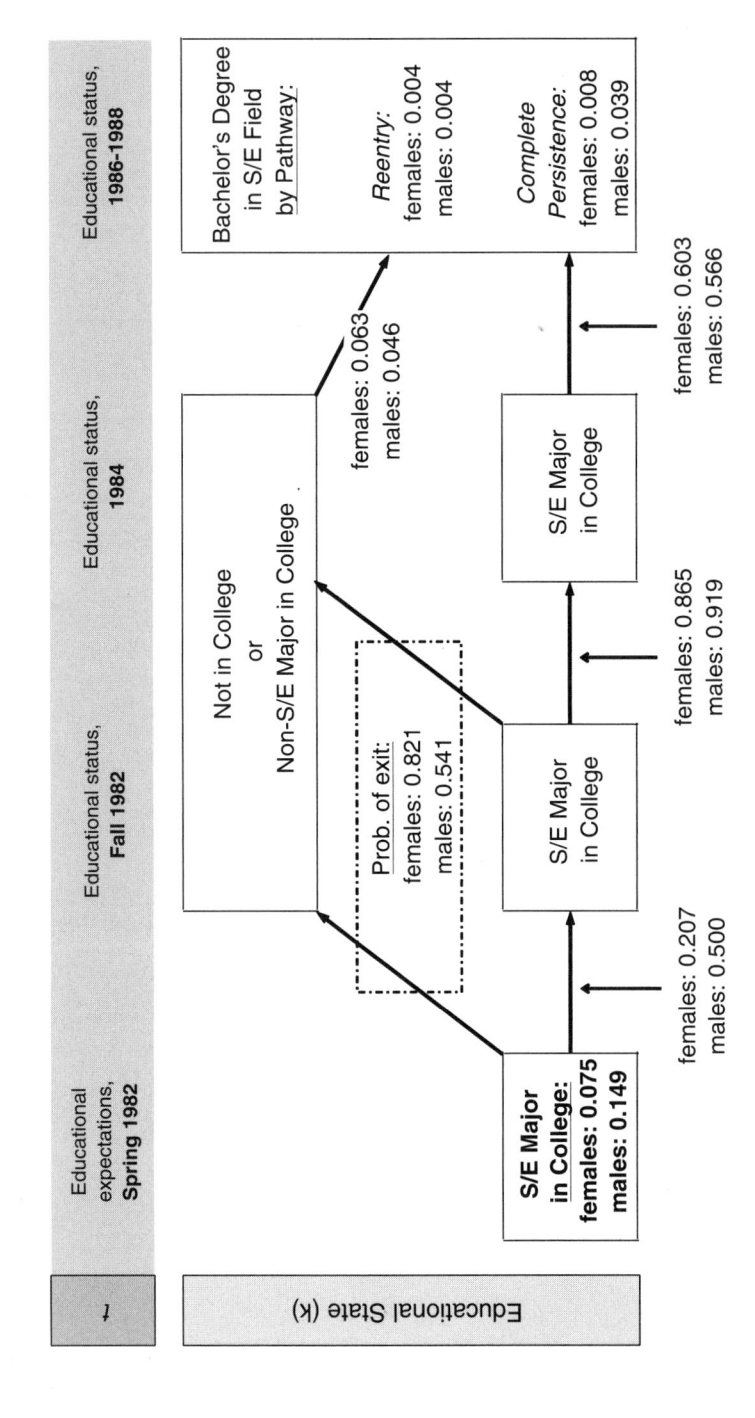

FIGURE 2-13 Sex-specific probabilities for selected pathways to an S/E baccalaureate.

In Figure 2-13, we observe that in the senior year of high school, women are much less likely than men to plan a science/engineering major in college. In addition, women experience a much larger attrition from the science/engineering educational trajectory than men do at the transition from high school to college. In the later college years, however, we find women and men to have similar transition rates to attaining degrees in science and engineering.

The "Productivity Puzzle"

In an influential paper, Cole and Zuckerman (1984) state that "women published slightly more than half (57%) as many papers as men." They found that the gender gap had persisted for many decades at this level and could not find any explanations for it. Out of despair, they called this gender difference the "productivity puzzle." Later, Long (1992), after considering possible explanations, reaffirms this characterization with the observation that "none of these explanations has been very successful."

We analyzed data from four nationally representative surveys of faculty in postsecondary institutions in 1969, 1973, 1988, and 1993.[3] Two major findings emerged from our work about the puzzle. First, sex differences in research productivity declined sharply between the 1960s and the 1990s, even without any controls. This is shown in Figure 2-14. Women scientists' research productivity has improved because their overall structural positions, such as institutional affiliation, have improved. This improvement in women's productivity relative to men's suggests that the large gender gap observed for earlier decades should not be attributed to innate biological differences between men and women.

Second, most of the observed sex differences in research productivity can be attributed to sex differences in background characteristics, employment positions and resources, and marital status. This is shown in Table 2-7. The first line of Table 2-7 reproduces the observed trends presented earlier in Figure 2-14. In lower lines, we included statistical adjustments for the fact that women and men differ in relevant characteristics, such as rank, year from a bachelor's degree to PhD, and institutional affiliation. Thus, even in the earlier decades, the observed sex differences in productivity can be explained once these relevant attributes are controlled for.

Family Life and Women Scientists' Careers

A common theme is the importance of considering the family in studies of women in science. In particular, we find that it is not marriage *per se* that hampers women's career development. Married women appear to be disadvantaged only if

---

[3]See also Y Xie and KA Shauman (1998). Sex differences in research productivity revisited: new evidence about an old puzzle. *American Sociological Review* 63:847-870.

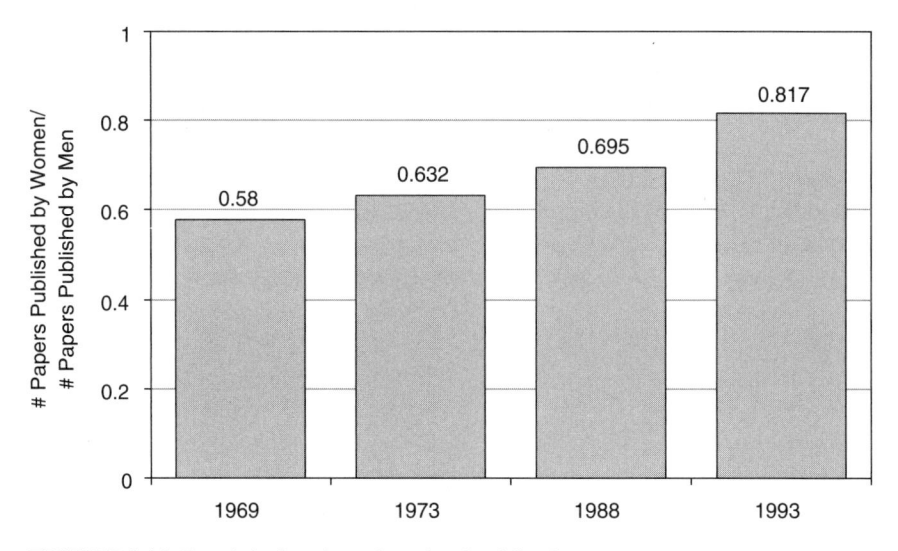

FIGURE 2-14 Trends in female-male ratio of publication rate.

TABLE 2-7 Estimated Female-to-Male Ratio of Publication

Model Description	1969	1973	1988	1993
(0): Sex	0.580***	0.632***	0.695**	0.817
(1): (0) + Field + Time for PhD + Experience	0.630***	0.663***	0.800	0.789*
(2): (1) + Institution + Rank +Teaching + Funding + RA	0.952	0.936	0.775	0.931
(3): (2) + Family/Marital Status	0.997	0.971	0.801	0.944

*$p<.05$ **$p<.01$ ***$p<.001$ (two-tailed test), for the hypothesis that there is no mean difference between males and females.

they have children. For example, we show that, relative to their male counterparts, married women with children are less likely to pursue careers in science and engineering after the completion of science/engineering education[4] less likely to be in the labor force or employed, less likely to be promoted,[5] and less likely to be geographically mobile.[6] Although some of the gender differences are attributable to the advantages that marriage and parenthood bestow upon men, they clearly suggest that being married and having children create career barriers that are unique to women—as opposed to men—scientists.

---

[4] Xie and Shauman (2003). Ibid, Chapters 5 and 6.
[5] Xie and Shauman (2003). Ibid, Chapter 7.
[6] Xie and Shauman (2003). Ibid, Chapter 8.

TABLE 2-8 Female-to-Male Odds Ratio of Post-Baccalaureate Career Paths by Family Status

Family Status	Grad School or Work	Grad School	Grad School in S/E	Work in S/E
Single	0.90	1.02	0.77	0.78**
Married without children	0.28***	0.67	0.11**	0.72**
Married with children	0.05***	0.35*		0.39***

*p<.05 **p<.01 ***p<.001 (two-tailed test), for the hypothesis that there is no mean difference between males and females.

Table 2-8 presents the female-to-male odds ratio of post-baccalaureate career paths by family status. There are five destinations for graduates with a bachelor's degree in science and engineering: (1) out of work and school altogether, (2) graduate school in science/engineering, (3) graduate school in nonscience/engineering, (4) work in science/engineering, and (5) work in nonscience/engineering. For the five outcomes, we made four contrasts and found that in all four, married women with children are disadvantaged in terms of science/engineering careers. Column 1 shows that married women with children are less likely than men to either work or attend graduate school. In column 2, we see that they are less likely than men to be in graduate school rather than working. Furthermore, married women with children are less likely than men to be in science/engineering, either in work (column 3) or in graduate school (column 4). Similarly, we also find married women with children disadvantaged in terms of other labor force outcomes.[7]

Summary

While the conventional wisdom often draws on casual analyses of nonrepresentative data, our tentative conclusions are based on very good data and careful analyses. Table 2-9 shows the contrast between conventional wisdom and our findings.

There appear to be two types of simplistic explanations. At one extreme, some observers claim that gender differences in science are all due to innate biological differences between men and women. At the other extreme, some scholars are tempted to make a sweeping claim that all gender differences are due to discrimination against women in school and at work. Our research shows that both positions are wrong. Otherwise, it would not be possible to explain either the rapid improvement of women's position in science, which cannot be attributed to

---

[7]Xie and Shauman (2003). Ibid, Chapter 7.

TABLE 2-9 Comparison Between Conventional Thinking and Our Findings

Conventional Wisdom	Our Findings
• Math deficiency • "Pipeline" paradigm • "War of the sexes" within marriage • Low rates of research productivity • Some "key" factor	• Gender gap in mathematics is small • Career processes are fluid and dynamic • Being married *and* having children matter • Sex differences in research productivity declined and can be attributed to differences in personal characteristics and structural features of employment • Deep social, cultural and economic roots

change in biological differences between the sexes, or the interaction between gender and parental status, which suggests that factors outside educational and work settings play an important role.

Women's underrepresentation in science/engineering has deep social, cultural, and economic roots that will not be transformed by a few isolated policy interventions or programs. Increasing women's representation in science/engineering requires many social, cultural, and economic changes that are large-scale and interdependent. After spending ten years searching for explanations, our research indicates we should stop looking for simple explanations and easy fixes, as attractive as they may be to us as human beings. Instead, we should look at the actual social processes that generate gender differences in science, and base policy interventions on empirical knowledge about these processes. Finally, while there may be policy changes that could address some of the complex reasons for women's underrepresentation, we should not expect any individual policy change to bring about gender equity in science overnight.

References

JR Cole and H Zuckerman (1984). The productivity puzzle: Persistence and change in patterns of publication of men and women scientists. *Advances in Motivation and Achievement* 2:217-258.
JS Long (1992). Measures of Sex Differences in Scientific Productivity. *Social Forces* 71:159-178.
LW Sells (1980). The mathematics filter and the education of women and minorities. In: *Women and the Mathematical Mystique*, eds. L Fox, L Brody, and D Tobin. Baltimore, MD: Johns Hopkins University Press.
Y Xie and KA Shauman (1998). Sex differences in research productivity revisited: New evidence about an old puzzle. *American Sociological Review* 63:847-870.
Y Xie and KA Shauman (2003). *Women in Science: Career Processes and Outcomes*. Cambridge, MA: Harvard University Press.

Section 3

Poster Abstracts

SOCIOLOGY

Florence Bonner and Vernese Edgeh, *Policy and Praxis: Advancing Women in Higher Education and Influencing Outcomes*

Miguel R. Olivas-Luján, Ann Gregory, John Miller, JoAnn Duffy, Suzy Fox, Terri Lituchy, Silvia Inés Monserrat, Betty Jane Punnett, and Neusa María Bastos F. Santos, *Successful Academic Women in the Americas: Human and Social Capital Descriptors*

Gloria Scott, *Science is Foundation for Leadership*

Roberta Spalter-Roth, *Work-Family Policies in Academia as Resources or Rewards*

Monica Young, *Case Studies from the Female Engineering Professoriate*

ORGANIZATIONAL STRUCTURE

Amber Barnato and Pamela Peele, *The Role of Informal Organizational Structures on Women in the Health Sciences*

Diana Bilimoria, Susan R. Perry, Xiangfen Liang, Patricia Higgins, Eleanor P. Stoller, and Cyrus C. Taylor, *How Do Female and Male Faculty Members Construct Job Satisfaction?*

Diana Bilimoria, C. Greer Jordan, and Susan R. Perry, *A Good Place to do Science: Creating and Sustaining a Productive, Inclusive Work Environment for Female and Male Scientists*

Diana Bilimoria, Margaret M. Hopkins, Deborah A. O'Neil, and Susan R. Perry, *An Integrated Coaching and Mentoring Program for University Transformation*

Cheryl Geisler, Deborah Kaminski, Robyn Berkley, and Linda Layne, *Up Against the Glass: Gender and Promotion at a Technological University*

Rachel Ivie, *Women in Academic Physics and Astronomy*

Mary Ellen Jackson, Phyllis Robinson, Sarah Conolly Hokenmaier, and J. Lynn Zimmer, *Faculty Horizons: Recruiting a Diverse Faculty*

Delia Saenz and Allecia Reid, *Diversity in STEM Disciplines: The Case of Faculty Women of Color*

INSTITUTIONAL POLICY

Ruth Dyer and Beth A. Montelone, *Initiatives to Increase Recruitment, Retention and Advancement of Women in Science and Engineering Disciplines at Kansas State University*

Lisa Frehill, Mary O'Connell, Elba Serrano, and Cecily Jeser-Cannavale, *Effective Practices for STEM Faculty Diversity*

Jo Handelsman, Molly Carnes, Jennifer Sheridan, Eve Fine, and Christine Pribbenow, *NSF ADVANCE at the UW-Madison: Three Success Stories*

Peggy Layne, Patricia Hyer, and Elizabeth Creamer, *Institutional Transformation at Virginia Tech*

Janet Malley, Pamela Raymond, and Abigail Stewart, *Institutional Transformation at the University of Michigan*

Nancy Martin, Beth Mitchneck, and William McCallum, *Scientifically Correct: Speaking to Scientists about Diversity*

Geraldine L. Richmond, *Working to Increase the Success of Women Scientists in Academia*

Eve A. Riskin, Kate Quinn, Joyce W. Yen, Sheila Edwards Lange, Suzanne Brainard, Ana Mari Cauce, and Denice D. Denton, *Leadership Workshops to Effect Cultural Change*

Tammy Smecker-Hane, Lisa Frehill, Priscilla Kehoe, Susan V. Bryant, Herb Killackey, and Debra Richardson, *ADVANCE: Successful Recruitment of Women to STEM at UCI*

POSTER ABSTRACTS

SOCIOLOGY

Policy and Praxis: Advancing Women in Higher Education and Influencing Outcomes

Florence Bonner and Vernese Edgeh, *Howard University*

Women in all parts of the world experience unequal playing fields in their quest for education, employment, occupational prestige, income and resources in nearly every discipline and field. Women remain heavily concentrated in the service fields in higher education and work. When we find more integration by gender men still occupy the positions with higher prestige, greater income and more resources. This is painfully so in the sciences. For example, in European Union (EU) countries such as Cyprus, Czech Republic, Estonia, Hungary, Latvia, Lithuania, Malta, Poland, Slovakia, and Slovenia, women represent the most students in service disciplines, education (74%), humanities and arts (66%), and health and welfare (72%). Men, on the other hand, comprise 77% of all students in the engineering, manufacturing, and construction fields (European Commission on Education, 2002). In South Africa, women graduates account for only 9% in engineering, 28% in agriculture, 38% in medicine, and 47% in the sciences. The most severe inequalities in South African higher education exist among African women (Government of South Africa, 1997).

In the U.S. as in many other countries of Europe, like France and the United Kingdom, women outnumber men in most institutions of higher education (Bonner, 2002), have higher grades upon entry and graduate faster; but men enter with more resources, more confidence (Allen 2005). On average men still outnumber women in most science fields and if they do not in the academy they do in the workplace. For example, data from the U.S. Bureau of Labor Statistics (2002) show that although women (14,621,158) outnumbered men (11,577,535) overall in the labor force, men held 2,218,400 positions in computer and mathematical occupations compared to 950,047 for women. Men held 86,343 positions in mathematical science occupations compared to women's 67,663. They dominated the architectural subcategories (2,301,953 men to 357,345 women) and in engineering fields (1,522,655 men to 179,800 women). These gender disparities prevail even in the academy in positions of power and authority; and in key places where mentoring routinely takes place.

We examine this problem within the context of the argument that— the presence of women in the academy in greater number than men, often with higher grades, faster time to graduation and success in graduation rates—women are just fine, it's men who are in trouble. Two questions focus the examination.

• Does a numerical majority in higher education entry and graduation rates constitute gender equality for women and does this numerical majority alone represent institutional change?
• Does an increased acquisition of advanced degrees translate into equity in outcomes such as employment, status, salary or resources?

Data compiled from the sources mentioned reveals that, for women, higher education achievement has not translated into gender equality within the academy or outside of it; rather, it has fueled an illusion and fostered a false premise of overwhelming success. Women still face many challenges inside institutions of higher education and learning as well as entry into nontraditional careers and professions; they have not reached parity with men nor have they surpassed them. Disproving the fallacy and debunking the myth that women have conquered all of the problems (or most) requires examination of at least the two questions above. We examine higher education success and outcomes, such as career choices of women and men; location (status and pay) in the occupational hierarchy and labor force to reflect on the questions in an effort to point to needed policy and support in the academy to remedy rather than exacerbate the conditions. We pay particular attention to African-American women.

Successful Academic Women in the Americas:
Human and Social Capital Descriptors

Miguel R. Olivas-Luján,[1] Ann Gregory,[2] John Miller,[3] JoAnn Duffy,[3] Suzy Fox,[4] Terri Lituchy,[5] Silvia Inés Monserrat,[6] Betty Jane Punnett,[7] and Neusa María Bastos F. Santos[8]

A complex interplay of personal and cultural characteristics enables some women, and not others, to overcome barriers to professional success. High-achieving women may share certain personal characteristics, beliefs, and experiences,

---

[1] Clarion University of Pennsylvania and ITESM, Monterrey Campus.
[2] Memorial University of Newfoundland.
[3] Sam Houston State University.
[4] Loyola University of Chicago.
[5] Concordia University.
[6] Universidad del Centro de Buenos Aires.
[7] University of the West Indies.
[8] Pontifica Catholica Universidade de Sao Paolo.

regardless of the countries in which they live. However, every individual is socialized within a particular national culture, and may be expected to share certain values and expectations with other members of that culture. The main goal of this research project was to identify similarities and differences across occupations (academic, professional, and managerial) of "successful" women in terms of personality, background and support structures, in various countries.

At the outset of the project, many facets of "success" were considered. For the purposes of this study, it was agreed to operationally define "success" as professional success, specifically "reaching a relatively high level in one's occupation or profession." The following criteria were used for participation in the study,—private sector managers of managers, academic tenured, full professors or senior university administrators, entrepreneurial women who have owned a business at least three years, government ministers/officials, and legal and medical professionals, as well as engineers.

Surveys and interviews were used to collect data on the following three sets of variables: National/Cultural (Collectivism/Individualism, Power Distance, Uncertainty Avoidance), Personal (Self-efficacy, Locus of Control, Need for Achievement), and Social-Experiential (Psychosocial and Career Mentoring). National/Cultural variables were measured using Dorfman & Howell's (1988) scales inspired by three of Hofstede's (1980) work dimensions. To measure personal variables, we used the following scales: Self efficacy was measured using an instrument developed by Sherer, Maddix, Mercandante, Prentice, Dunn, Jacobs, & Rogers (1982). A work related locus of control scale derived by Spector (1988) from Rotter's (1960) work was used to measure the extent to which one perceives being in control of events in one's life. Need for achievement scale was drawn from the Jackson's (1989) Personality Research Form. Finally, Psychosocial and Career Mentoring was measured with a scale by Tepper, Shaffer, & Tepper (1996).

Over 1,100 professionally successful women and 531 undergraduate business students completed the above surveys. In addition, researchers completed semi-structured interviews with a minimum of 25 participants in each of the countries. The international team, led by eight researchers from diverse academic perspectives (management, strategy, history, women's studies, human resources, and organization behavior) focused on the following countries or regions: United States, Canada, Mexico, Brazil, Argentina, Chile, and the English-speaking Caribbean.

This presentation will compare the subset of academic women in the sample with other sub-samples from the study. Findings will be discussed in the context of the Convocation.

Science is Foundation for Leadership

Gloria Scott, *Jarvis Christian College*

In academia, science is one of the major "subject matter areas"—humanities, social sciences, science representing the basic educational core of knowledge. Science students must internalize and utilize the "scientific method" which is fundamental. One of the basic reasons that science is required as a part of the educational core, is that the exposure to knowledge acquisition and utilization within a methodology forms a foundation for all intellectual interchange and exploration. As one reviews the data on women leaders in the United States, especially African American women, significant numbers were science majors and worked in teaching, and research. Hundreds of them lead in the broad non-profit sector, in the educational non non-profit sector as well as in the profit sector. They are present as executive and volunteer leaders at local, state national and international levels. This relationship is essential to communicate inter-generationally to current and future college women to help them understand the professional foundation and implementation that science provides as an occupational area, but also to know that science foundation knowledge and experience provides a complex interplay with creating self assured, high performance leadership ability. Science represents the most important fundamental source of knowledge, analysis, strategy and understanding to facilitate human achievement in organizational frames. The poster presents this relationship as essential and foundational in the production of leaders.

Work-Family Policies in Academia as Resources or Rewards

Roberta Spalter-Roth, *American Sociological Association*

There is a growing broad-based, social movement to ameliorate the time conflict between work and family by increasing the availability of work family-policies to academic faculty. This movement responds to the growing numbers of women PhDs in the sciences and other disciplines, and the failure of these women to attain the highest ranks at research universities. Pressure from this movement has expanded the range of institutions of higher education have begun to offer at least minimal work-family policy options so that women (and men) can reconcile the demands of two "greedy institutions." Two sorts of arguments are made to bring about change: (1) needs based or resources policy and (2) "best and the brightest" or rewards policy. To test these arguments, we analyze evidence from a survey of sociology PhDs, 6 years after they obtained their PhDs. We find that academic mothers who use of at least one work-family policy significantly increase their scholarly productivity, in the form of peer-reviewed publications, without increasing their time spent at work. These findings suggest that over-all, work-family policies may be effective in meeting the demands of both greedy institutions. Yet, these policies, including family leave, extending the tenure clock, modified teaching loads, and part-time tenure track positions do not appear to be distributed as resources to all academic mothers with young children. Rather they appear to be distributed as rewards on the basis of the predicted productivity of faculty mothers. Predicted productivity is measured by the prestige of the graduate school attended and the publications completed in graduate school. These findings suggest that chairs and other administrators may be less willing to distribute resources to mothers who are not perceived as the "best and the brightest." To make these policies more universal, needs-based policies, chairs need to inform themselves about the entitlement to work-family policies, deans need to hold chairs accountable for their distribution, and provosts need to hold deans accountable. The broad-based, multi-organizational social movement supporting work-family policies needs to continue to monitor institutions of higher education.

Case Studies from the Female Engineering Professoriate

Monica Young, *Syracuse University*

This research study focused on a desire to understand the reasons why women enter the engineering profession, as well as how they succeed in this profession and ultimately become members of the engineering professoriate. As a female engineer who changed fields, I had a range of experiences both good and bad that contributed to my decision. I am passionate about my current field, science education, and I wish to recruit more females into science and engineering by working in this field. The goal for this study was to find women who have found success in engineering, and question them about the aspects of their lives that helped them succeed. Two women in academia who hold doctoral degrees in engineering were the cases for this study. The women were selected based on differences in their backgrounds, both academically and personally. Each woman was interviewed extensively to garner information about her experiences in elementary and secondary school, college, graduate school, and life in general. Some of the general themes that emerged throughout these interviews were: how to make science memorable, the role of mentoring, the importance of questioning, and social norms. The women discussed experiences they had throughout their academic career that contributed to their current success as assistant professor and senior administrator. Though much information was gleaned from analyzing the interviews, there is a great deal more to learn from these women. Future research will further question the participants in this study and expand the number of participants.

ORGANIZATIONAL STRUCTURE

The Role of Informal Organizational Structures on Women in the Health Sciences

Amber Barnato and Pamela Peele, *University of Pittsburgh*

Women in academic science careers often confront organizational structures developed to foster success among men. While these organizational structures may function well for men, they do not necessarily serve well the objectives of recruiting, hiring, retaining, and promoting the careers of women in science careers. Ample social research documents that women and men differ along many domains including their risk preferences, their career choices, and social interactions. Given this, it should not be surprising that the formal organizational structures developed to promote the success of men in academia are not optimal structures for women. We report on the impact of overlaying of informal organizational structures onto the standard organizational structure of academia on the recruitment, hiring, retention, success, and well-being of professional women in the health sciences. We implemented an informal structure that consisted of a core group of junior women health services research faculty at the University of Pittsburgh. This started with a group of three junior women faculty in 1997. From that group, it has grown to over 20 women in the health sciences across the University, most hired after the implementation of the core group. The informal structures in place provide women with a feeling of belonging and friendship which is an important aspect for the recruiting of new women. There is a robust information exchange over such topics as diverse as childcare resources and contract negotiations that allows women to easily observe the experiences of other women and to avoid common pitfalls facing junior women in health sciences. The core group provides several important functions including the endowment of new members with professional capital. An important development of this informal structure is a snowball effect that has produced several new auxiliary social groups that specialize in a variety of topics such as cooking clubs, book clubs, working mom clubs, etc. Each group is informally attached to the core group of research women and while the groups overlap to some extend, they are closed sets. The result is that as the informal structure evolves and expands, it creates mutations to serve the current needs of women in the health sciences while still preserving the core group. The informal structure has served to recruit, hire, and retain women in the health sciences, an effect that grows with the increasing robustness of the structure itself. Two of the most important elements of the informal structure include the rapid access to information and the championing of each other's work. With a single e-mail request, women can activate the informal group to find necessary information from a nanny to accompany them to a conference so they can present their work to information on how someone negotiated their last contract. By the same mechanism, women in the group seem to have a high propensity to promote the work of others in the group. We are now beginning to apply some qualitative methods to investigate what the core elements are that allowed this mechanism to be successful when attempts by others to this have failed.

How Do Female and Male Faculty Members Construct Job Satisfaction?

Diana Bilimoria, Susan R. Perry, Xiangfen Liang, Patricia Higgins, Eleanor P. Stoller, and Cyrus C. Taylor, *Case Western Reserve University*

In this study we examine how a sample of 248 male and female professors at a Midwestern private research university construct their academic job satisfaction. Our findings indicate that both women and men perceive that their job satisfaction is influenced by the institutional leadership and mentoring they receive, but only as mediated by the two key academic processes of access to internal academic resources (including research-supportive workloads) and internal relational supports from a collegial and inclusive immediate work environment. Gender differences emerged in the strengths of the perceived paths leading to satisfaction: women's job satisfaction derived more from their perceptions of the internal relational supports than the academic resources they received whereas men's job satisfaction resulted equally from their perceptions of internal academic resources and internal relational supports received. Implications for leadership and institutional practices are drawn from the findings.

A Good Place to Do Science: Creating and Sustaining a Productive, Inclusive Work Environment for Female and Male Scientists

Diana Bilimoria, C. Greer Jordan, and Susan R. Perry, *Case Western Reserve University*

The purpose of our study was to identify and better understand the work environment factors that lead to the development, retention, and advancement of women faculty in a university setting. Thus, we conducted a case study of a top-ranked science department in a Tier 1 research university. The department, whose primary faculty consisted of three female and thirteen male scientists, had achieved a reputation for cooperation, advancement of women, and productive outcomes. Over a six-month period, we collected data using multiple qualitative methods including interviews, direct observation, and archival research. Inductive analysis of this data revealed five overarching factors and 12 subfactors that contributed to the cooperative, inclusive, productive work culture. The five overarching factors include a shared scientific identity; constructive interactions;

participative department activities, inclusive department subprocesses and integrative leadership practices. We tapped existing literature to synthesize these factors into a process model of an inclusive, productive work culture. This study integrates several theoretical approaches to creating effective, diverse work groups into one model. Our work also highlights the role of member identity and types of interactions in building inclusive, high performing work groups across demographic differences. The findings also have implications for intervening in groups, departments, or teams as part of efforts to attract and retain a broader range of high quality scientists, including women and minorities.

An Integrated Coaching and Mentoring Program for University Transformation

Diana Bilimoria, Margaret M. Hopkins,
Deborah A. O'Neil, and Susan R. Perry,
Case Western Reserve University and University of Toledo

Higher education researchers and university administrators alike are increasingly concerned about the persistent dearth of women faculty, the overall glacial advancement of women, and the existence of a glass ceiling in academic science and engineering fields. The sources of these problems may be traced to individual psychological processes (gender schemas) and systematic institutional barriers, resulting in perceptions of a chilly climate for women scientists and engineers in academia (Sandler and Hall, 1986), the experience of subtle discrimination by women faculty (Blakemore, Switzer, DiLorio, and Fairchild, 1997), the slow but steady accumulation of disadvantage over the course of women's academic careers (Valian, 1999), and the flight from academia by women scientists and engineers at every step in the educational pipeline.

Today, leading universities are beginning to undertake comprehensive remedies to address these problematic attitudinal and structural issues. Prominent within the approaches being implemented are a variety of coaching and mentoring initiatives aimed at helping women faculty succeed, particularly in the early and middle stages of their careers, and at helping key upper- and mid-level university leaders (deans and chairs) in changing the culture of their academic units. We believe that the combined focus of short term coaching targeted at empowering personal and professional development together with long term mentoring and sponsorship can help women faculty succeed in academia. Targeted coaching initiatives designed to assist academic decision makers such as deans and department chairs in understanding their roles in creating inclusive, supportive environments can also help curb the leaky pipeline of faculty women in sciences and engineering. In this report we describe the activities, challenges, and successes of a unique multi-level, integrated coaching and mentoring initiative at our university.

Up Against the Glass: Gender and Promotion at a Technological University

Cheryl Geisler, Deborah Kaminski, Robyn Berkley, and Linda Layne,
Rensselaer Polytechnic Institute

Despite increasing access to faculty ranks, women faculty members continue to encounter a glass ceiling when it comes to achieving the rank of full professor. At Rensselaer, we have been engaged in a research program aimed at documenting, understanding, and changing such differential patterns of advancement. Our work began with the development of a low-cost metric, the 13+ Club Index that can be used to monitor advancement in institutions and organizations. The 13+ Club Index examines the ratio between the percentage of women are 13 or more years past degree and have not yet been promoted to full professor and the percentage of men in the same situation. If the women and men at an institution in the 13+ Club are being promoted at the same rate, this index will be 1.

Our first project showed how this index can be used to monitor and change patterns of differential advancement. In particular, a study of the promotion patterns at Rensselaer completed in 2002 showed that women with 13 or more years since highest degree were 2.2 times than men more likely to remain unpromoted to the rank of full professor. Subsequent to the distribution of the results of this study, numerous changes, both institutional and individual, took place. As a consequence, by the time of our next analysis, two and one-half years later, 5 of the 11 women who had not been promoted in the original analysis had gone up for and received promotion. Overall, the rate of promotion for women at Rensselaer was more than three times the rate for men and the number of women full professors on the faculty doubled.

Our second project sought to understand the processes underlying differential patterns of advancement. A stratified sample of associate and full professors matched by school and gender were surveyed. Based on this data, we developed six profiles, and found that the distribution of men and women over these profiles was quite distinct. First, looking just at those who had been promoted to full professor, we found that women were more likely to fit Profile III (promoted to full after denial and with no advice or encouragement), while men were more likely to fit either Profile I or II (promoted on first try). Second, looking at those who had not been promoted, we found men were more likely to fit Profile IV (not seeking a promotion to full despite advice and encouragement), while women were more likely to fit Profile V (not seeking promotion nor were they advised or encouraged). Finally, we broke down the entire sample in the 13+ group based on advice and encouragement and found 8 of 11 males were advised and/or encouraged to go up for promotion, however, only 4 of 12 women were so advised.

Our research suggests three forces combine to challenge institutions working to improve women's advancement. To begin with, it appears that whenever the climate at an institution improves with respect to advancement, men will benefit as well as women. Inequities between men and women can thus remain despite improvements in women's situations. Next, pipeline issues are notoriously difficult to ameliorate. While it may be possible to reduce the rate of nonpromotion among women relatively quickly, reducing the flow of the pipeline into the ranks of the nonpromoted may be a longer term project. And finally, achieving equity in senior hires is particularly difficult. While processes can be put into place to insure a diverse pool of applicants, the pool of available women applicants at the senior rank is still limited.

Women in Academic Physics and Astronomy

Rachel Ivie, *American Institute of Physics*

One characteristic of the structure of physics and astronomy departments is that the representation of women decreases with each step up the academic ladder. Although women are about half of high school physics students, they make up less than one-fourth of physics bachelor's degree recipients. Women earn about 18% of PhDs in physics, but comprise only 10% of the faculty. At stand-alone astronomy departments, 14% of the faculty members are women, even though women earn 26% of astronomy PhDs. In spite of this apparent leak in the pipeline, our data show that women are represented on physics and astronomy faculties at levels consistent with degree production in the past. In addition, there are only small differences in the dropout rate for male and female physics graduate students. Our data show that there are a few physics departments that have done an outstanding job in recruiting and retaining women faculty and students. There are also serious problems related to the structure of academic employment. For example, women physicists are hired as instructors and adjuncts at rates greater than they are hired into ranked faculty positions. The reasons for this disparity are unknown, but should be investigated.

Faculty Horizons: Recruiting a Diverse Faculty

Mary Ellen Jackson, Phyllis Robinson, Sarah Conolly Hokenmaier, and
J. Lynn Zimmer
ADVANCE Program, University of Maryland, Baltimore County

The underrepresentation of women faculty in science, technology, engineering, and mathematics (STEM) fields is a longstanding national problem. A 2005 study shows that female faculty in the top 50 research universities are underrepresented at all ranks, especially as full professors. The study also points out that underrepresented minority women "are almost nonexistent in science and engineering departments at research universities" and are less likely than Caucasian women or men of any race to be awarded tenure or reach full professor status (Nelson and Rogers, 2005). The University of Maryland, Baltimore County (UMBC), a research university committed to excellence and inclusiveness, received an Institutional Transformation Award from the National Science Foundation's ADVANCE Program to address these issues. As part of this program, UMBC created *Faculty Horizons*, a two-day workshop focused on postdoctoral research fellows and upper level graduate students, particularly women in STEM fields, to provide these future faculty with the knowledge and tools necessary to build a successful career. In recognizing the national problem of the severe shortage of women from underrepresented groups in STEM, special attention is paid to including African American and Hispanic women.

Diversity in STEM Disciplines: The Case of Faculty Women of Color

Delia Saenz and Allecia Reid, *Arizona State University*

Structural, dynamic, and social factors preclude women from equal status, representation, and empowerment in STEM disciplines across the country. The confluence of racial/ethnic minority status and gender, and their concomitant impact, further exacerbate the lack of full participation and recognition of underrepresented women of color in these fields. The presentation will elucidate social psychological factors such as tokenism, stereotypy, and confirmation bias that play a role in inhibiting capacity among women scientists, in general, and women of color scientists, in particular. Research findings from an ongoing cohort study, funded by the Ford Foundation, will be presented. The research involved interviews, focus groups, and Web surveys at approximately 20 of the top PhD-producing, public, research extensive universities in our nation. Specifically, the research questions focused on institutional climate as perceived by both women faculty themselves and by institutional officials (provost, general counsel, affirmative action officers). In addition to providing comparative analyses of these varied institutional citizen perspectives, the data include examples of factors, initiatives, and practices that facilitate/inhibit inclusive excellence. The presentation will further identify critical forces at different levels of university functioning (individual, unit, institutional culture) that affect outcomes for STEM faculty. Some of these factors parallel those faced by underrepresented members of the academy across non-STEM disciplinary fields. Other factors appear to be unique to the STEM disciplines. Challenges and opportunities associated with differential levels of institutional diversification will be addressed. Finally, recommendations for 'best practices' that can be implemented at different levels of institutional functioning will be suggested. Among these are strategies that women belonging to both mainstream and minority populations can engage to promote their own success; cultural adaptations that can be implemented within departments and colleges; and policies and procedures, along with leadership imperatives that must be in place to achieve transformational outcomes. A model of interdependence will be invoked to conceptualize the current gaps in the academy, potential interventions (including educational programs for all faculty, staff, and administrators), and identification of critical goals for institutions of higher education, particularly in their role of inspiring knowledge acquisition and dissemination in the service of producing an educated citizenry. The significance of these needed changes stems not only from a current capacity perspective within STEM fields, but also from the reality of the student and workforce pipelines, and from the critical need to ensure national and global technological progress.

INSTITUTIONAL POLICY

Initiatives to Increase Recruitment, Retention and Advancement of Women in Science and Engineering Disciplines at Kansas State University

Ruth Dyer and Beth A. Montelone, *Kansas State University*

Kansas State University (K-State) has implemented a number of programs over the last ten years designed to increase the success of women in science and engineering (S&E) disciplines. These programs address issues pertinent to beginning, mid-career, and senior faculty members. One of these is the KSU Mentoring Program for Women and Minorities in the Sciences and Engineering. It has been in existence since 1993, supported by funding from the Sloan Foundation and the K-State Office of the Provost. It is a competitive program that pairs untenured faculty members with mentors in their research areas and provides small awards (up to $6000) that can be used in a variety of ways. To date, 52 individuals have received awards; ten of these individuals are women of color and five are men of color. The tenure success rate of the 28 individuals who have become eligible for tenure is 79%, higher than the average rate for both men and women in S&E departments and university-wide. 18 of the 22 faculty members receiving tenure are still at K-State, and five women are already full professors. An analysis conducted in 2002 of 31 recipients of the Mentoring Awards indicated that these faculty members had at that time generated over 500 publications, 15 other pieces of intellectual property, and over $39 million in extramural grant funds.

In 2003, K-State received an ADVANCE Institutional Transformation Award from the National Science Foundation. Our project includes initiatives for individual departments and colleges, as well as project-wide programs, to improve recruitment, retention, and advancement of women in S&E. In the first two years of our project, we have made sixty professional development awards to women faculty members to facilitate their participation in professional conferences, collaboration with colleagues at other institutions, and initiation of research projects. Six tenured women faculty members have received awards to enhance their research activities or undertake administrative projects through interaction with senior mentors. Eighteen untenured women have hosted leaders in their disciplines as part of the ADVANCE Distinguished Lecture Series. Furthermore, 20 men and women faculty members in the College of Veterinary Medicine have established two peer mentoring groups that provide a series of activities to enhance professional development. Departments in the College of Engineering may propose novel strategies for effective recruitment of women; two departments received funding for this purpose in 2004-2005 and successfully hired three women faculty members. Moreover, eight additional women were hired into tenure-track positions in other S&E departments in 2004-2005. This is more than double the average number of women hired in S&E departments over the last 6 years. Further, six women scientists or engineers have been appointed to administrative positions (Department Head, Associate Dean, Associate Provost) since the start of the project. We believe that these recent hires and administrative appointments reflect an increased commitment to the inclusion and advancement of women in S&E at K-State. We are encouraged by the success of these programs but recognize that continued progress requires constant scrutiny and sustained diligence.

Effective Practices for STEM Faculty Diversity

Lisa Frehill, Mary O'Connell, Elba Serrano, and Cecily Jeser-Cannavale
University of California, Irvine and New Mexico State University

What role do department chairs and deans play in ensuring diversity within academe? This presentation is the culmination of a year of work by a diverse group of 40 deans, department chairs/heads, and senior faculty. After attending conferences with programming about diversity in the professoriate program participants attended a three-day writing retreat. The culmination of this effort are several products on one CD: the Dean's Guide to Diversity, the Department Chair's/Head's Guide to Diversity, and a set of PowerPoint presentation slides that could be used by faculty and academic administrators to convince their peers of the merits of engaging in various "best practices" to increase faculty diversity. While many other excellent guides to diversity have been published, these products feature the "voice" of faculty and academic administrators who have actually implemented and worked with the practices suggested by others. Elements of the publications will be presented on the poster.

NSF ADVANCE at the UW-Madison: Three Success Stories

Jo Handelsman, Molly Carnes, Jennifer Sheridan, Eve Fine, and Christine Pribbenow
University of Wisconsin-Madison

In this poster, we highlight—the hiring process, work/life balance, and departmental climate. We introduce three new initiatives funded by the NSF ADVANCE Institutional Transformation award designed to address these problem areas on the UW-Madison campus. We describe our efforts to raise awareness of how unconscious biases might impact hiring by training chairs of hiring committees; we outline our Life Cycle Research Grant program which provides research funds to faculty who are experiencing a life event that impacts their research productivity; and we outline our workshops for department chairs and the process we use to help them improve the climate in their departments. We present evaluation data indicating the effectiveness of the programs, and show progress of institutionalization and dissemination of the programs.

Institutional Transformation at Virginia Tech

Peggy Layne, Patricia Hyer, and Elizabeth Creamer, *Virginia Tech*

Virginia Tech is one of 19 recipients of a five-year, $3.5 million, institutional transformation grant from the National Science Foundation's ADVANCE program to increase the participation and success of women faculty in science and engineering. Now in its third year, AdvanceVT is taking a multifaceted approach to change at Virginia Tech. Activities include preparing women graduate students in science and engineering for faculty careers, working with search committees to help them understand and address unintended bias in the hiring process and to develop diverse candidate pools for faculty positions, providing untenured women faculty with research seed money to help them develop more competitive proposals for external funding, developing leadership skills to enable tenured women faculty to take on leadership roles in the university, building community among women across departments and colleges, raising awareness of gender issues among university leaders, and reviewing, revising, and overseeing implementation of university policies that disproportionately impact women faculty. Throughout the program, AdvanceVT is collecting data on career aspirations and job satisfaction of both male and female faculty at Virginia Tech and tracking statistics on the numbers of women at all levels at the institution. This poster will highlight AdvanceVT program activities, impacts, and plans for sustainability beyond the grant period.

Institutional Transformation at the University of Michigan

Janet Malley, Pamela Raymond, and Abigail Stewart, *University of Michigan*

The NSF ADVANCE Project at the University of Michigan (UM ADVANCE), housed within the Institute for Research on Women and Gender, is a five-year, grant funded project to promote institutional transformation in science and engineering fields by increasing the participation, success, and leadership of women faculty in academic science and engineering.

Initiatives to support individual women scientists and engineers include faculty career advising, research funds, and a network of women scientists and engineers. The Elizabeth C. Crosby and Lydia A. DeWitt Research Funds were established to help meet career-relevant needs of individual instructional track faculty and research track faculty, respectively, if meeting those needs will help increase the retention or promotion of women scientists and engineers. The Network of Women Scientists and Engineers, which is composed of tenure-track women faculty in science and engineering departments across the entire campus, meets several times each year to socialize, to talk about issues the members have in common, and to develop plans for the future. A number of UM ADVANCE activities—many of the leadership development activities, the mentoring initiatives, the annual report to the campus about our progress—have emerged from Network discussions.

UM ADVANCE also provides support to departments aiming to improve their climates through transformation grants, self-studies, and reviews. It encompasses initiatives at all levels of the University, including data-based workshops presented by the Science and Technology Recruiting to Improve Diversity and Excellence Committee (STRIDE) and interactive theater performances by the CRLT Players. More specifically, the STRIDE Committee provides information and advice about practices that will maximize the likelihood that well-qualified female and minority candidates for faculty positions will be identified, and, if selected for offers, recruited, retained, and promoted at the University of Michigan. The committee works with departments by meeting with department chairs, faculty search committees, and other departmental leaders involved with recruitment and retention. The CRLT Players have developed three ADVANCE sketches focusing on mentoring, faculty hiring, and the tenure decision process. These performances are based on faculty interviews and focus groups conducted at the University of Michigan. The performances demonstrate the challenges female faculty may encounter in interactions with other faculty and provide a foundation for dialogue about climate and collegiality.

The President and Provost set in motion a comprehensive review of University policies that affect women scientists and engineers. As co-chairs of the Gender in Science and Engineering Committee (GSE), the President and Provost charged three subcommittees (in turn chaired by three deans), to examine policies in three areas: faculty evaluation and development; recruitment, retention and leadership; and family policies and faculty tracks. This initiative began a process of institutionalizing practices that will be useful for both male and female faculty, while focusing on the policies that research shows affect women more, such as family-related policies, the tenure clock, and the criteria for evaluation and promotion.

Scientifically Correct: Speaking to Scientists about Diversity

Nancy Martin, Beth Mitchneck, and William McCallum, *University of Arizona*

The University of Arizona is currently developing a program to train trainers to orient search committees about scientific research on how unconscious bias can influence the search and hiring processes. This effort is part of a larger National Sciences Foundation ADVANCE proposal (currently under review). Other ADVANCE institutions (University of Michigan, University or Wisconsin at Madison, and others) have used search training and recruitment teams successfully. We extend this by tailoring the orientation materials to specific colleges and developing a cohort of male and female faculty to deliver the message. Our strategy is to reach scientists by sharing the latest and best social science research on unconscious bias. This evidence comes primarily from the field of social psychology, and includes both laboratory and field experiments. Our training provides research evidence of bias on the part of well-intentioned actors. Importantly, unconscious gender bias occurs in both women and men. We provide practical strategies supported by additional research evidence for overcoming the problem of unconscious bias. Also under development are toolkits for interviewing and conducting hiring negotiations.

Working to Increase the Success of Women Scientists in Academia

Geraldine L. Richmond, *University of Oregon*

As scientists, we leave graduate school with a toolbox full of skills to help us to design and conduct scientific experiments, analyze data, publish papers, and to communicate scientific concepts to others. Unfortunately, this toolbox often does not include skills that enable us to communicate effectively in a variety of professional settings or negotiate for what we need in order to successfully achieve our career goals.

In this poster I describe some of the workshops available to women graduate students, postdocs and faculty around the country that teach such skills. These workshops have been developed by COACh (Committee on the Advancement of

Women Chemists) and have been shown to be highly effective in helping women to advance in their careers and reduce the stress in their personal lives (Richmond, 2005) The full day workshops have been designed to (1) enhance communication and negotiation skills needed for effective teaching and career development, (2) teach leadership techniques that are effective for women scientists in an academic setting (3) provide a forum for networking with other academic women scientists and engineers and (4) develop effective strategies for making institutional and departmental change that improves the climate, recruiting and retention of underrepresented groups. Case studies, theatre, role-playing and lively debate contribute to the learning experience for the 15-20 participants in each session. COACh also offers workshops for minority women scientists and engineers that address these above described issues while also providing a forum for discussion of how these methods can be effectively used to address problems of a racial nature that are faced by women in these populations. The workshop facilitators are experienced professional women in human resources, leadership training, teaching, and higher education administration, with extensive experiences in many professional venues.

Over 1000 women scientists and engineers from academic institutions across the country have thus far participated in these workshops. Our research on the impact of these workshops on participants shows that they are significantly enhancing their career progress. New workshops and forums are being launched that are specifically targeted towards institutional transformation. Descriptions of these workshops and information on how to bring them to your professional meeting or institution can be found on the COACh Web site: http://coach.uoregon.edu/.

COACh was formed in 1998 by a group of women professors in the chemical sciences concerned about the slow progress of women in their profession and its impact the ability to attract and retain younger female talent into the field. Details about other COACh activities can be found on the Web site. COACh is grateful for funding from the National Science Foundation (Chemistry and the ADVANCE program), the National Institutes of Health, and Basic Energy Sciences from the Department of Energy.

Leadership Workshops to Effect Cultural Change

Eve A. Riskin, Kate Quinn, Joyce W. Yen, Sheila Edwards Lange, Suzanne Brainard, Ana Mari Cauce, and Denice D. Denton, *University of Washington, ADVANCE Center for Institutional Change*

Institutional transformation as intended by the NSF ADVANCE program requires a significant amount of change in attitudes, practices and policies throughout the university community. The success of institutional change hinges largely on the extent to which change occurs at the academic department level

(Bennett and Figuli, 1990; Lucas, 2000). Yet, academic department chairs are not often prepared to be change agents or administrative managers (Lucas, 2000; Gmelch and Miskin, 1995; Wolverton, Gmelch, Montez, and Nies, 2001). Faculty who have risen to the department chair position are usually recognized leaders in their scholarly fields and have been trained to be scholars, not managers. Most come to the department chair position with little leadership training beyond leading departmental committees (Seagren, Cresswell, and Wheeler, 1993). Department chair orientation and training, if provided, is often once a year and limited to administrative and fiscal responsibilities which represent the tip of the iceberg of a department chair's responsibilities. Often, the more challenging and rewarding experiences of department chairs relate to mentoring faculty and managing their concerns. Gmelch & Miskin found that the responsibilities that chairs rate as most important (i.e. the recruitment and selection of faculty, the evaluation of faculty performance, conflict resolution and leadership) are absent from orientations and campus-based training programs. And while department chairs may seek guidance from online and printed resources targeted at department chairs, such resources are generally not campus-specific enough to be sufficient.

As part of its institutional change efforts, the UW ADVANCE program sought to provide department chairs with ongoing opportunities to draw from the experience and wisdom of their department chair colleagues and to conscientiously explore topics relevant to equity in science and engineering and the success of their faculty and departments. Each academic quarter, the CIC hosts a half-day leadership workshop for department chairs, deans, and emerging leaders. These workshops serve as a forum for cross-college networking and professional development for chairs and emerging leaders and are designed to engage academic leaders as critical actors in changing institutional culture. Prior to this program, department chairs received little or no professional development beyond their initial orientation to the department chair position. Evaluations of these workshops have been uniformly high, and department chairs have stated these workshops are the "boot camp" they never got. This poster provides an overview of the quarterly leadership workshop program, offers recommendations for replication, and discusses results from two national workshops modeled after the quarterly workshop program.

ADVANCE: Successful Recruitment of Women to STEM at UCI

Tammy Smecker-Hane, Lisa Frehill, Priscilla Kehoe, Susan V. Bryant, Herb Killackey and Debra Richardson,
University of California, Irvine

The NSF-funded ADVANCE: Institutional Transformation Program at the University of California at Irvine has two significant and lasting innovations related to increasing faculty diversity, which will be covered in the presentation. First, since September 2002 each of the university's ten schools has had at least one "Equity Advisor," who serves as a faculty advisor to the school's dean on issues related to gender equity. Equity Advisors meet with the dean, search committees, department chairs and other faculty in their respective schools to raise awareness and use of more proactive search strategies to increase recruitment of women to the tenure-track faculty ranks. Equity Advisors also a run faculty mentoring programs for newly hired assistant professors. A related innovation is the use of a series of three university forms in the faculty search process that documents the use of proactive strategies and ensures the transparency of search processes. These forms are titled "Search Plan and Advertisement for Regular Ranks Faculty," "Interim Search Activities Statement," and "Final Search Activities Statement." The second of these three forms is new for the 2005-2006 academic year. All three of these forms require an Equity Advisor signature, which increases search transparency and oversight related to equity issues in each search at the UCI.

Section 4

Appendixes

Appendix A   Workshop Agenda

Appendix B   Speaker Biographical Information

Appendix C   Committee Biographical Information

Appendix D   Statement of Task

APPENDIX A

The National Academies
Committee on Science, Engineering, and Public Policy

Committee on Maximizing the Potential of Women in Academe
CONVOCATION ON BIOLOGICAL, SOCIAL, AND ORGANIZATIONAL CONTRIBUTIONS TO SCIENCE AND ENGINEERING SUCCESS

December 9, 2005

National Academy of Sciences Building
2101 Constitution Avenue, NW
Washington, DC

AGENDA

9:00 Welcome
Wm. A. Wulf, President, National Academy of Engineering

9:05 Keynote: *Factors that Determine Success in Science and Engineering Careers*
Donna Shalala [IOM], Chair, Committee on Maximizing the Potential of Women in Academe

9:45 Plenary Discussion 1: *Cognitive and Biological Contributions*
Moderator: Ana Mari Cauce, member, Committee on Maximizing the Potential of Women in Academe

- Gender similarities
  Janet Hyde, Department of Psychology, University of Wisconsin-Madison
- Sexual dimorphism in the developing brain
  Jay Giedd, National Institute of Mental Health, NIH
- Environment-genetic interactions in the adult brain: effects of stress on learning
  Bruce McEwen [NAS/IOM], The Rockefeller University
- Biopsychosocial contributions to cognitive performance
  Diane Halpern, Berger Institute for Work, Family, and Children, Claremont McKenna College

11:15 Break

11:30 **Plenary Discussion 2:** *Social Contributions*
Moderator: Alice Agogino, member, Committee on Maximizing the Potential of Women in Academe

- Implicit and explicit gender discrimination
  Mahzarin Rustum Banaji, Department of Psychology, Harvard University, and Radcliffe Institute for Advanced Study
- Contextual influences on performance
  Toni Schmader, Department of Psychology, University of Arizona
- Interactions between power and gender
  Susan Fiske, Department of Psychology, Princeton University
- Social influences on science and engineering career decisions
  Yu Xie, Department of Sociology, University of Michigan

1:00 **Lunch** *Poster Session in the Great Hall*

2:00 **Plenary Discussion 3:** *Organizational Structures*
Moderator: Lotte Bailyn, member, Committee on Maximizing the Potential of Women in Academe

- Competence assumptions and stereotype-driven evaluations
  Joan Williams, Center for WorkLife Law, University of California, Hastings College of the Law
- Economics of gendered distribution of resources in academe
  Donna Ginther, Department of Economics, University of Kansas
- The value of work-family policies
  Robert Drago, Departments of Labor and Women's Studies, Pennsylvania State University
- Gendered organizations
  Joanne Martin, Graduate School of Business, Stanford University

3:15 **Break**

3:30 **Plenary Discussion 4:** *Implementing Policies*
Moderator: Nan Keohane, member, Committee on Maximizing the Potential of Women in Academe

- Recruitment practices
  Angelica Stacy, Department of Chemistry, University of California, Berkeley
- Reaching into minority populations
  Joan Reede, Harvard Medical School

- Creating an inclusive work environment
  Sue Rosser, Ivan Allen College, Georgia Tech
- Successful practices in industry
  Kellee Noonan, Diversity Program Manager, Technical Career Path, Hewlett Packard

4:45   **Plenary Discussion 5:** *Open Q&A with Committee*

5:30   **Closing Comments**
       Denice Denton, Member, Committee on Maximizing the Potential of Women in Academe

5:45   **Reception in Great Hall**

6:30   **Adjourn**

Copies of the presentations will be available shortly after the Convocation at http://www7.nationalacademies.org/womeninacademe/Convocation.html.

APPENDIX B
SPEAKER BIOGRAPHICAL INFORMATION

Mahzarin Banaji is Richard Clarke Cabot Professor of Social Ethics in the Department of Psychology at Harvard University and Carol K. Pforzheimer Professor at the Radcliffe Institute for Advanced Study. She was born and raised in India, in the town of Secunderabad, where she attended St. Ann's High School. Her BA is from Nizam College in Hyderabad and her MA in psychology from Osmania University. She received her PhD from Ohio State University (1986), was a postdoctoral fellow at University of Washington, and taught at Yale University from 1986 until 2001 where she was Reuben Post Halleck Professor of Psychology. In 2002 she moved to Harvard University. Banaji studies human thinking and feeling as it unfolds in social context. Her focus is primarily on thinking and feeling systems that operate in implicit or unconscious mode. In particular, she is interested in the unconscious nature of assessments of self and other humans that reflect feelings and knowledge (often unintended) about their social group membership (e.g., age, race/ethnicity, gender, class). From such study of attitudes and beliefs of adults and children, she asks about the social consequences of unintended thought and feeling. Her work relies on cognitive/affective behavioral measures and neuroimaging (fMRI) with which she explores the implications of her work for theories of individual responsibility and social justice. Banaji is a Fellow of the American Association for the Advancement of Science, of the American Psychological Association and of the American Psychological Society. She served as Secretary of the APS, on the Board of Scientific Affairs of the APA, and on the Executive Committee of the Society of Experimental Social Psychology. She has served as Associate Editor of Psychological Review and of Journal of Experimental Social Psychology and is currently Co-Editor of Essays in Social Psychology. She serves on the editorial board of several journals, among them *Psychological Science, Psychological Review, Journal of Personality and Social Psychology*, and *The DuBois Review*. Banaji was Director of Undergraduate Studies at Yale for several years, chaired APS's Task force on Dissemination of Psychological Science, and served on APA's Committee on the Conduct of Internet Research. Among her awards, she has received Yale's Lex Hixon Prize for Teaching Excellence, a James McKeen Cattell Fund Award, and fellowships from the Guggenheim Foundation, and the Rockefeller Foundation's Bellagio Study Center. In 2000, her work with R. Bhaskar received the Gordon Allport Prize for Intergroup Relations. With Anthony Greenwald and Brian Nosek, she maintains an educational website that has accumulated over 2.5 million completed tasks measuring automatic attitudes and beliefs involving self, other individuals, and social groups. It can be reached at http://www.implicit.harvard.edu.

Robert Drago is a Professor of Labor Studies and Women's Studies at the Pennsylvania State University. He is also Professorial Fellow at the University of Melbourne and moderates the work/family newsgroup on the internet (lsir.la.psu.edu/workfam). He holds a PhD in Economics from the University of Massachusetts at Amherst, and has been a Senior Fulbright Research Scholar. Drago's recent research concerns biases against caregiving in the workplace, working time, the value of work-family policies. He also studies college and university faculty and public policies related to work and family with funding from the Alfred P. Sloan Foundation. Most recently, in conjunction with Jackie Rogers and Theresa Vescio, he completed research on the relative decline of women as intercollegiate coaches, with funding from the NCAA and NACWAA. He is president elect for 2006 of the College and University Work/Family Association, a cofounder of the Take Care Net, the 2001 recipient of the R.I. Downing Fellowship from the University of Melbourne (Australia), serves on the board of the Berger Institute for Work, Family and Children, is a member of the Council on Contemporary Families and the International Association for Feminist Economics, and serves on the advisory board for the Ms. Foundation's *Take Our Daughters and Sons to Work* day. He has published numerous articles in publications such as *Academe, American Behavioral Scientist, Handbook of Work and Family, Industrial and Labor Relations, Journal of Labor Economics, and the Monthly Labor Review.*

Susan T. Fiske is professor of psychology at Princeton University. She has taught on the faculties of the University of Massachusetts, Amherst, and Carnegie Mellon University. A 1978 Harvard PhD, she received an honorary doctorate from the Université Catholique de Louvain, Louvain-la-Neuve, Belgium, in 1995. Her graduate text with Shelley Taylor, *Social Cognition* (1984; 2nd ed., 1991), defined the subfield of how people think about and make sense of other people. Her 2004 text, *Social Beings: A Core Motives Approach to Social Psychology*, describes people's most relevant evolutionary niche as social groups, with core motives (such as belonging) that enable people to adapt. Her research has focused on how people choose between category-based (stereotypic) and individuating impressions of other people, as a function of power and interdependence. Her current research shows that social structure predicts distinct kinds of bias against different groups in society, some more disrespected and others more disliked. Her expert testimony in discrimination cases includes one cited by the U.S. Supreme Court in a 1989 landmark case on gender bias. In 1998, she also testified before President Clinton's Race Initiative Advisory Board. Fiske won the 1991 American Psychological Association Award for Distinguished Contributions to Psychology in the Public Interest, Early Career, in part for the expert testimony. She also won, with Glick, the 1995 Allport Intergroup Relations Award from the Society for the Psychological Study of Social Issues for work on ambivalent sexism. Among other elected offices, Fiske was president of the American Psychological Society for 2002–2003. She edited, with Daniel Gilbert and Gardner Lindzey, the *Handbook of Social Psychology* (4th ed., 1998) and with Daniel Schacter and Carolyn Zahn-Waxler, the *Annual Review of Psychology* (Vols. 51-60, 2000-2009). She has served on the boards of Scientific Affairs for the American Psychological Association, the American Psychological Society, Annual Reviews Inc., the Social Science Research Council, and the Common School in Amherst.

Jay Giedd is the Chief of the HtmlResAnchor Unit on Brain Imaging in the Child Psychiatry Branch at the NIMH. He received his MD from the University of North Dakota in 1986, training in adult psychiatry at the Menninger Foundation in Topeka, KS, and Child and Adolescent Psychiatry training at Duke University in Durham, NC. He is board certified in General, Child and Adolescent, and Geriatric Psychiatry. His research focuses on the relationship between genes, brain, and behavior in healthy development and in neuropsychiatric disorders of childhood onset. His laboratory is conducting longitudinal neuropsychological and brain imaging studies of healthy twins and singletons as well as clinical groups such as Attention-Deficit/Hyperactivity Disorder, Childhood-Onset Schizophrenia, and others. Over the past 10 years they have acquired over 3000 MRI scans making this the largest pediatric neuroimaging project of its kind. The lab also studies sexual dimorphism in the developing brain, especially important in child psychiatry where nearly all disorders have different ages of onsets, prevalence, and symptomatology between boys and girls, by exploring clinical populations which have unusual levels of hormones (e.g., Congenital Adrenal Hyperplasia, Familial Precocious Puberty) or variations in the sex chromosomes (e.g., Klinefelter's Syndrome, XYY, XXYY). The lab is also conducting studies of monozygotic and dizygotic twins which are beginning to unravel the relative contributions of genes and environment on a variety of developmental trajectories in the pediatric brain. The group is also involved in the development and application of techniques to analyze brain images and is actively collaborating with other imaging centers throughout the world to advance the image analysis field.

Donna Ginther is an Associate Professor of Economics at the University of Kansas. Prior to joining the University of Kansas faculty, she was a research economist and associate policy adviser in the regional group of the Research Department of the Federal Reserve Bank of Atlanta. From 1997 to 2000, she was an assistant professor of economics at Washington University, and from 1995 to 1997 she was an assistant professor of economics at Southern Methodist University. Her major fields of study are scientific labor markets, gender differences in employment outcomes, wage inequality, and children's educational attainments. Ginther has been published in several journals, including the *Journal of the American Statistical Association, Journal of Economic Perspectives, Demography,* and the *Papers and Proceedings of the American Economic Association.* She is a member of the American Economics Association and the Population Association of America. As of 2006, she is a member of the Board of the Committee on the Status of Women in the Economics Profession of the American Economic Association. A native of Wisconsin, Ginther received her doctorate in economics in 1995, master's degree in economics in 1991, and bachelor of arts in economics in 1987, all from the University of Wisconsin-Madison.

Diane F. Halpern is Professor of Psychology and Director of the Berger Institute for Work, Family, and Children at Claremont McKenna College. She is the past-president (2005) of the American Psychological Association. Halpern has won many awards for her teaching and research, including the 2002 Outstanding Professor Award from the Western Psychological Association, the 1999 American Psychological Foundation Award for Distinguished Teaching, 1996 Distinguished Career Award for Contributions to Education given by the American Psychological Association, the California State University's State-Wide Outstanding Professor Award, the Outstanding Alumna Award from the University of Cincinnati, the Silver Medal Award from the Council for the Advancement and Support of Education, the Wang Family Excellence Award, and the G. Stanley Hall Lecture Award from the American Psychological Association. She is the author of many books: *Thought and Knowledge: An Introduction to Critical Thinking; Thinking Critically About Critical Thinking* (with Heidi Riggio), *Sex Differences in Cognitive Abilities; Enhancing Thinking Skills in the Sciences and Mathematics, Changing College Classrooms; Student Outcomes Assessment*; and *States of Mind: American and Post-Soviet Perspectives on Contemporary Issues in Psychology* (coedited with Alexander Voiskounsky). Her most recent book is co-edited with Susan Murphy, entitled *From Work-Family Balance to Work-Family Interaction: Changing the Metaphor*.

Janet Hyde is Helen Thompson Woolley Professor of Psychology and Women's Studies at the University of Wisconsin-Madison. She earned her PhD in 1972 from the University of California, Berkeley. She is the author of a textbook for the psychology of women course, entitled *Half the Human Experience: The Psychology of Women*. One line of her research has focused on gender differences in abilities and self-esteem. Another line focuses on women, work, and dual-earner couples. One current research project, the Wisconsin Maternity Leave and Health Project (now called the Wisconsin Study of Families and Work), focuses on working mothers and their children; this research has public policy implications in the area of parental leave. Another current project, funded by the National Science Foundation, is the Moms & Math (M&M) Project, in which she is studying mothers interacting with their 5th or 7th grade children as they do mathematics homework together. Other research investigates gender differences in the emergence of depression and negative cognitive style in adolescence. She is a fellow of the American Psychological Association and the American Association for the Advancement of Science, and a winner of the Heritage Award from the Society for the Psychology of Women for career contributions to research on the psychology of women and gender.

Joanne Martin is the Fred H. Merrill Professor of Organizational Behavior and, by courtesy, Sociology at the Graduate School of Business, Stanford University. Martin received a PhD in Social Psychology from Harvard in 1977 and honorary doctorates from Copenhagen Business School in 2001 and the Vrej University in Amsterdam in 2005. Her current research focuses on gender in organizations, including subtle barriers to advancement for women and how to structure gender equity change programs. She is also known for her research on organizational culture (books include *Cultures in Organizations: Three Perspectives* and *Organizational Culture: Mapping the Terrain*). She was elected to serve on the Board of Governors of the Academy of Management and the Faculty Advisory Board (seven elected members) at Stanford University. She also has been a member of the Board of Directors of C.P.P., Inc., where she was the lead outside director, and the International Advisory Board of the International Center for Research in Organizational Discourse, Strategy, and Change, for the Universities of Melbourne, Sydney, London, and McGill. Martin has received numerous awards, including the Gordon Allport Intergroup Relations Award from the American Psychological Association in 1988 (for a paper with Thomas Pettigrew on barriers to inclusion for African-Americans); the Distinguished Educator Award from the Academy of Management in 2000 (for doctoral education), the Centennial Medal from the Graduate School of Arts and Sciences, Harvard University, for research-based contributions to society, in 2002; and the Distinguished Scholar Career Achievement Award from the National Academy of Management, Organization and Management Theory Division, in 2005.

Bruce McEwen [NAS/IOM] is the Alfred E. Mirsky Professor and Head of the Harold and Margaret Milliken Hatch Laboratory of Neuroendocrinology at The Rockefeller University. McEwen graduated Summa Cum Laude in Chemistry from Oberlin College in 1959 and obtained his PhD in Cell Biology in 1964 from The Rockefeller University. He returned to Rockefeller in 1966 to work with the psychologist, Neal Miller, after postdoctoral studies in neurobiology in Sweden and a brief period on the faculty at the University of Minnesota. He was appointed as Professor at Rockefeller in 1981. He is a member of the US National Academy of Sciences, the Institute of Medicine, the American Academy of Arts and Sciences and a Fellow of the New York Academy of Sciences. He served as Dean of Graduate Studies from 1991-1993 and as President of the Society for Neuroscience in 1997-1998. As a neuroscientist and neuroendocrinologist, McEwen studies environmentally-regulated, variable gene expression in brain mediated by circulating steroid hormones and endogenous neurotransmitters in relation to brain sexual differentiation and the actions of sex, stress and thyroid hormones on the adult brain. His laboratory discovered adrenal steroid receptors in the hippocampus in 1968. His laboratory combines molecular, anatomical, pharmacological, physiological and behavioral methodologies and relates their findings to human clinical information. He is a member of the MacArthur Foundation Research Network on Socioeconomic Status and Health, in which he is helping to reformulate concepts and measurements related to stress and stress hormones in the context of human societies. He is the co-author with science writer Elizabeth Lasley of the book for a lay audience called *The End of Stress as We Know It* published by the Joseph Henry Press and the Dana Press (2002).

Kellee Noonan is a manager for the development of HP technical women worldwide and in that context, is the Diversity Program Manager for the Hewlett Packard Technical Career Path. The program was initiated by the CTO and implemented 2 years ago to shatter the glass ceiling for individual contributor technologists and allow them a non-management career path up to executive levels. The goal of the program is to help HP attract, retain, challenge, and engage the world's strongest technical talent at all levels of the company. Noonan received her MS in Mechanical Engineering Design from Stanford University, and her BS in Mechanical Engineering from the University of the Pacific in Stockton, CA. At HP, Noonan has held a variety of positions including R&D engineer, Program Manager for HP Corporate Continuing Engineering Education, Computer Systems Technical Education Manager, and an Organizational Effectiveness Consultant. Prior to HP, Noonan was a Member of Technical Staff at the Jet Propulsion Laboratories in Pasadena, California.

Joan Reede is the Dean for Diversity and Community Partnership at Harvard Medical School where she works to recruit and prepare minority students for jobs in the biomedical professions, and to promote better health care policies for the benefit of minority populations. She is the first African American woman to hold that rank at Harvard Medical School and one of the few African American women to hold a deanship at a medical school in the United States. She earned her BS from Brown in 1977 and her MD from Mt. Sinai School of Medicine in 1980. She completed an internship and residency in pediatrics at Johns Hopkins University School of Medicine and a fellowship in child psychiatry at Boston's Children's Hospital. Thereafter, she went on to earn two more degrees, an MPH in 1990 and an MS in 1992 from Harvard School of Public Health. At Harvard, Dr. Reede was struck by the absence of minorities among the School of Public Health faculty. In 1990, after a year as a fellow at Harvard Medical School, Reede and several colleagues founded the Biomedical Science Careers Program (BSCP), to match minority students from high school through post-graduate levels with mentors in their fields of interest. Dr. Reede is also a founder and director of the Commonwealth Fund/Harvard University Fellowship in Minority Health Policy, which offers physicians with an interest in minority and disadvantaged populations a year of professional training for leadership positions in health care policy and practice.

Sue Rosser has served as Dean of Ivan Allen College, the liberal arts college at Georgia Institute of Technology, since 1999; she is also Professor of History, Technology, and Society. She received her PhD in Zoology from the University of Wisconsin-Madison in 1973. From 1995-1999, she was Director for the Center for Women's Studies and Gender Research and Professor of Anthropology at the University of Florida-Gainesville. In 1995, she was Senior Program Officer for Women's Programs at the National Science Foundation. From 1986 to 1995 she served as Director of Women's Studies at the University of South Carolina, where she also was a Professor of Family and Preventive Medicine in the Medical School. She has edited collections and written approximately 115 journal articles on the theoretical and applied problems of women and science and women's health. She is author of the books *Teaching Science and Health from a Feminist Perspective: A Practical Guide* (1986), *Feminism within the Science and Health Care Professions: Overcoming Resistance* (1988), *Female-Friendly Science* (1990), *Feminism and Biology: A Dynamic Interaction* (1992), *Women's Health: Missing from U.S. Medicine* (1994), and *Teaching the Majority* (1995), *Re-engineering Female Friendly Science* (1997), and *Women, Science, and Society: The Crucial Union* (2000). Her latest book is *The Science Glass Ceiling: Struggles of Academic Women Scientists* (2004). She also served as the Latin and North American co-editor of *Women's Studies International Forum* from 1989-1993 and currently serves on the editorial boards of *NWSA Journal*, *Journal of Women and Minorities in Science and Engineering* and *Women's Studies Quarterly*. She has held several grants from the National Science Foundation, including "A USC System Model for Transformation of Science and Math Teaching to Reach Women in Varied Campus Settings" and "POWRE Workshop"; she currently serves as co-PI on a $3.7 million ADVANCE grant from NSF. During the fall of 1993, she was Visiting Distinguished Professor for the University of Wisconsin System Women in Science Project.

Toni Schmader is an Associate Professor of Psychology at the University of Arizona. She received a BA with Honors, *summa cum laude* from Washington & Jefferson College and a PhD in Social Psychology from the University of California, Santa Barbara. Her research seeks to understand the interplay between self and social identity, particularly when one's social identity is accorded lower status or is targeted by negative group stereotypes. In exploring these issues, her research draws upon and extends existing theory on social stigma, social justice, social cognition, intergroup emotion, self-esteem, and motivation and performance. Her work has received funding from the National Science Foundation and the National Institute of Mental Health. In 2000, she was awarded the *Social Issues Dissertation Award* from the *Society for the Psychological Study of Social Issues* for her research examining how social status and the perceived legitimacy of that status influence the domains that people value. Her more recent research explores the impact of gender stereotypes on women's involvement and performance in math related domains.

Angelica Stacy is a Professor of Chemistry and Associate Vice Provost for Faculty Equity at the University of California at Berkeley. She received her Ph.D. at Cornell University (1981) and was a Postdoctoral Fellow at Northwestern University (1981-1983). From there she moved to Cornell University where she was an assistant professor in the Department of Chemistry. She moved to UC Berkeley in 1988. Interest in the Stacy Lab is in solid-state inorganic chemistry, with particular emphasis on the synthesis and characterization of new solid state materials with novel electronic and magnetic properties. Stacy was a National Science Foundation Presidential Young Investigator awardee (1984-1989). She has received a number of teaching and research excellence awards, including the Prytanean Society Faculty Enrichment Award, 1986; Exxon Fellowship for Solid State Chemistry, 1987; Sloan Foundation Fellowship (1988-1990); Camille and Henry Dreyfus Teacher-Scholar Award (1988); Distinguished Teaching Award, University of California (1991), Faculty Award for Women Scientists and Engineers, National Science Foundation (1991); Lawrence Berkeley Laboratory Technology Transfer Certification of Merit (1991); President's Chair for Teaching, University of California (1993-1996); Francis P. Garvan-John M. Olin Medal, American Chemical Society (1994), Catalyst Award, Chemical Manufacturers Association (1995); The Donald Sterling Noyce Prize for Excellence in Undergraduate Teaching (1996); Iota Sigma Pi Award for Professional Excellence (1996); and James Flack Norris Award for Outstanding Achievement in the Teaching of Chemistry (1998).

Joan C. Williams is Distinguished Professor of Law and Founding Director of the Center for WorkLife Law at University of California, Hastings College of the Law. A prize-winning author and expert on work/family issues, she is author of *Unbending Gender: Why Family and Work Conflict and What To Do About It* (Oxford University Press, 2000), which won the 2000 Gustavus Myers Outstanding Book Award. She has authored or co-authored four books and over fifty law review articles; her work is reprinted in casebooks on six different subjects; she has given over two hundred speeches and presentations in North and Latin America to groups as diverse as the National Employment Lawyers' Association, the Denver Rotary Club, the American Philosophical Association, and the Modern Language Association, and has lectured at virtually every leading U.S. university. Founding Director of WorkLife Law (WLL), she is also co-director of the Project on Attorney Retention. She has played a leading role in documenting workplace bias against mothers. Her "Beyond the Maternal Wall: Relief for Family Caregivers Who are Discriminated against on the Job," 26 *Harvard Women's Law Review* 77 (2003), (co-authored with Nancy Segal), was prominently cited in *Back v. Hastings on Hudson Union Free School District*, 2004 U.S. App. Lexis 6684 (2d Cir. April 7, 2004). She also has played a central role in organizing social scientists to document maternal wall bias, notably in a special issue of the *Journal of Social Issues* (2004), co-edited with Monica Biernat and Faye Crosby, which was awarded the Distinguished Publication Award by the Association for Women in Psychology. Her current work focuses on social psychology, and on how work/family conflict affects families across the social spectrum, with a particular focus on how caregiving issues arise in union arbitrations. For more information visit www.worklifelaw.org and www.pardc.org. Williams teaches property as well as courses related to gender, family and employment. She has two children. Her husband is a public interest lawyer specializing in privacy and internet issues.

Yu Xie is the Otis Dudley Duncan Professor of Sociology in the Department of Sociology at the University of Michigan. He is also affiliated with the Department of Statistics, the Population Studies Center, and the Survey Research Center of the Institute for Social Research and the Center for Chinese Studies. Yu Xie has a PhD in Sociology from the University of Wisconsin-Madison, an MA in the History of Science and an MS in Sociology both from the University of Wisconsin-Madison. He received a BS in Metallurgical Engineering from Shanghai University of Technology (1982). Yu Xie's main areas of interest are social stratification, demography, statistical methods, and sociology of science. He is the co-author of the recent book, *Women in Science, Career Processes and Outcomes*, which won the 2005 Choice Magazine Outstanding Academic Title. He has served as the Deputy Editor of *American Sociological Review* (1996-2000), the Associate Editor of the *Journal of the American Statistical Association* (1999-2001), member of several editorial boards, advisory panel member for the Sociology Program (1995-1997) and the Methodology, Measurements, and Statistics Program (2004-2006) at the National Science Foundation. He has held several distinguished faculty positions including assistant professor (1989-1994), associate professor (1994-1996), and professor (1996-present) in the Department of Sociology at the University of Michigan. Yu Xie is the recipient of numerous awards including the National Academy of Education Spencer Fellowship (1991-1992), the National Science Foundation's Young Investigator Award (1992-1997), the William T. Grant Foundation's Faculty Scholar Award (1994-1999), and the John Simon Guggenheim Memorial Foundation Fellowship (2002-2003). He received the Academician recognition from the Taiwan Academia Sinica (1994), and the American Academy of Arts and Science Fellow Award (2004). In addition, he has received several Teaching awards From University of Michigan including the Teaching Development Award from the Center for Research on Learning and Teaching (1990-1991), as well as the Excellence in Education Award from the College of Literature, Science, and the Arts (1992).

APPENDIX C
COMMITTEE ON MAXIMIZING THE POTENTIAL OF WOMEN IN ACADEMIC SCIENCE AND ENGINEERING

Biographical Information

DONNA E. SHALALA (CHAIR) became Professor of Political Science and President of the University of Miami on June 1, 2001. Born in Cleveland, Ohio, President Shalala received her AB degree in history from Western College for Women and her PhD degree from The Maxwell School of Citizenship and Public Affairs at Syracuse University. A leading scholar on the political economy of state and local governments, she has also held tenured professorships at Columbia University, the City University of New York (CUNY), and the University of Wisconsin-Madison. She served in the Carter administration as Assistant Secretary for Policy Development and Research at the US Department of Housing and Urban Development. From 1980 to 1987 she served as president of Hunter College of the City University of New York, and from 1987 to 1993 was Chancellor of the University of Wisconsin-Madison. In 1993 President Clinton appointed her U.S. Secretary of Health and Human Services (HHS) where she served for eight years, becoming the longest serving HHS Secretary in US history. At the beginning of her tenure, HHS had a budget of nearly $600 billion, which included a wide variety of programs including Social Security, Medicare, Medicaid, Child Care and Head Start, Welfare, the Public Health Service, the National Institutes of Health (NIH), the Centers for Disease Control and Prevention (CDC), and the Food and Drug Administration (FDA). President Shalala has more than three-dozen honorary degrees and a host of other honors. In 1992, *Business Week* named her one of the top five managers in higher education. She also received the 1992 National Public Service Award, and the 1994 *Glamour* magazine Woman of the Year Award. In 2005 she was named one of "America's Best Leaders" by US News and World Report and the Center for Leadership at Harvard University's Kennedy School of Government. She has been elected to the Council on Foreign Relations, National Academy of Education, the National Academy of Public Administration, the American Academy of Arts and Sciences, the National Academy of Social Insurance, the American Academy of Political and Social Science, and the Institute of Medicine of the National Academy of Sciences.

ALICE M. AGOGINO is the Roscoe and Elizabeth Hughes Professor of Mechanical Engineering and affiliated faculty at the University of California, Berkeley Haas School of Business in their Operations and Information Technology Management Group. She directs the Berkeley Expert Systems Technology (BEST) Laboratory and the Berkeley Instructional Technology Studio (BITS). She is currently Vice Chair of the Berkeley Division of the Academic Senate and served as Chair during the 2005-2006 academic year. She has served in a number of administrative positions at UC Berkeley including Associate Dean of Engineering and Faculty Assistant to the Executive Vice Chancellor and Provost in Educational Development and Technology. She also served as Director for Synthesis, an NSF-sponsored coalition of eight universities with the goal of reforming undergraduate engineering education, and continues as PI for the NEEDS and the digital libraries of courseware in science, mathematics, engineering and technology. She has supervised 65 MS projects/theses, 26 doctoral dissertations and numerous undergraduate researchers. Agogino is a registered Professional Mechanical Engineer in California and is engaged in a number of collaborative projects with industry. Prior to joining the faculty at UC Berkeley, she worked in industry for Dow Chemical, General Electric and SRI International. Her research interests include intelligent learning systems; information retrieval and data mining; multiobjective and strategic product design; nonlinear optimization; probabilistic modeling; intelligent control and manufacturing; sensor validation, fusion and diagnostics; wireless sensor networks; multimedia and computer-aided design; design databases; design theory and methods; MEMS synthesis and computer-aided design; artificial intelligence and decision and expert systems; and gender equity. She is a member of AAAI, AAAS, ACM, ASEE, ASME, AWIS, IEEE, NAE and SWE. She serves on the editorial board of three professional journals and has provided service on a number of governmental, professional, and industry advisory committees. Agogino received a BS in Mechanical Engineering from the University of New Mexico (1975), MS degree in Mechanical Engineering (1978) from the University of California at Berkeley, and PhD from the Department of Engineering-Economic Systems at Stanford University (1984). She received an NSF Presidential Young Investigator Award in 1985. She is an AAAS Fellow, is a member of the National Academy of Engineering and the European Academy of Science; is a Fellow of the Association of Women in Science; and was awarded the NSF Director's Award for Distinguished Teaching Scholars in 2004.

LOTTE BAILYN is a Professor of Management (in the Organization Studies Group) at MIT's Sloan School of Management and Co-Director of the MIT Workplace Center. In her work she has set out the hypothesis that by challenging the assumptions in which current work practices are embedded, it is possible to meet the goals of both business productivity and employees' family and community concerns, and to do so in ways that are equitable for men and women. Her most recent book, *Beyond Work-Family Balance: Advancing Gender Equity and Workplace Performance* with Rhona Rapoport, Joyce K. Fletcher, and Bettye H. Pruitt (Jossey Bass, 2002) chronicles a decade of experience working with organizations that supports this hypothesis, while also showing how difficult it is to challenge workplace assumptions. She currently serves on the National Academies Committee on Women in Science and Engineering.

ROBERT J. BIRGENEAU became the ninth chancellor of the University of California, Berkeley, on September 22, 2004. An internationally distinguished physicist, he is a leader in higher education and is well known for his commitment to diversity and equity in the academic community. Before coming to Berkeley, Birgeneau served 4 years as president of the University of Toronto. He previously was dean of the School of Science at the Massachusetts Institute of Technology, where he spent 25 years on the faculty. He is a foreign associate of the National Academy of Sciences, has received many awards for teaching and research, and is one of the most cited physicists in the world for his work on the fundamental properties of materials. A Toronto native, Birgeneau received his BSc in mathematics from the University of Toronto in 1963 and his PhD in physics from Yale University in 1966. He served on the faculty of Yale for one year, spent one year at Oxford University, and was a member of the technical staff at Bell Laboratories from 1968 to 1975. He joined the physics faculty at MIT in 1975 and was named chair of the physics department in 1988 and dean of science in 1991. He became the 14th president of the University of Toronto on July 1, 2000. At Berkeley, Birgeneau holds a faculty appointment in the Department of Physics in addition to serving as chancellor.

ANA MARI CAUCE is the Executive Vice Provost and Earl R. Carlson Professor of Psychology, University of Washington. She graduated from Yale University, earning a PhD in Psychology in 1984. She began teaching at the University of Washington in 1986 in the Department of Psychology. She also has a joint appointment in the Department of American Ethnic Studies and an adjunct appointment in Women's Studies. Cauce currently holds the Earl R. Carlson Professorship in Psychology and is Chair of the Department of Psychology. Since she began her graduate work, Cauce has been particularly interested in normative and non-normative development in ethnic minority youth and in at-risk youth more generally. She has published almost a hundred articles and chapters and has been recipient of grants from the W.T. Grant Foundation, the National Institute of Mental Health, the National Institute of Child Health and Human Development, and the National Institute of Alcoholism, and Alcohol Abuse. She is the recipient of numerous awards, including recognition from the American Psychological Association for Excellence in Research on Minority Issues; Distinguished Contribution Awards from the Society for Community Research and Action; and the American Psychological Association Minority Fellowship program. She has also received the University of Washington's Distinguished Teaching Award. Cauce is currently President-Elect of the Society for Community Research and Action.

CATHERINE D. DEANGELIS is Editor-in-Chief of The Journal of the American Medical Association, Editor-in-Chief of Scientific Publications and Multimedia Applications, and Professor of Pediatrics, Johns Hopkins University School of Medicine. She received her MD from the University of Pittsburgh's School of

Medicine, and her MPH from the Harvard Graduate School of Public Health (Health Services Administration), and her pediatric specialty training at the Johns Hopkins Hospital. DeAngelis oversees *JAMA* as well as nine *Archives* publications and *JAMA* related Web site content. Before her appointment with *JAMA*, she was vice dean for Academic Affairs and Faculty at Johns Hopkins University School of Medicine, and from 1994-2000, she was editor of *Archives of Pediatrics and Adolescent Medicine*. She also has been a member of numerous journal editorial boards. She has authored or edited 11 books on Pediatrics and Medical Education and has published more than 200 original articles, chapters, editorials, and abstracts. Most of her recent publications have focused on conflicts of interest in medicine, on women in medicine, and on medical education. Dr. DeAngelis is a council member of the Institute of Medicine of the National Academy of Sciences, a Fellow of the American Association for the Advancement of Science, and has served as an officer of numerous national academic societies including past chairman of the American Board of Pediatrics and Chair of the Pediatric Accreditation Council for Residency Review Committee of the American Council on Graduate Medical Education.

DENICE DENTON (deceased) was the Chancellor at the University of California, Santa Cruz. She formerly served as Dean of and a professor at the University of Washington's College of Engineering since 1996. Prior to her appointment as dean in 1996, she was a faculty member in electrical engineering and chemistry at the University of Wisconsin-Madison. While at the University of Washington, Denton led the development of the Faculty Recruitment Toolkit, a resource for attracting a top notch and diverse faculty. In a single year (2001) nine faculty members received the prestigious NSF Career Award. In addition, federal research funding more than doubled in 3 years (1998-2001), from $33.1 million in grants and contract awards to more than $75 million. She also emphasized implementing effective ways to teach a diverse engineering student body using a more project-oriented, experiential approach. This is facilitated by the Center for Engineering Learning and Teaching (CELT), the first center of its kind when established in 1998. She directed the University of Washington's NSF ADVANCE program for advancing women faculty in science and engineering. In 2004 Denton was honored by the White House with the Presidential Award for Excellence in Science, Mathematics and Engineering Mentoring, recognizing her role as a national leader in engineering education. Denton chaired the National Academy of Engineering's Board on Engineering Education from 1996 to 1999. She was a fellow of the American Association for the Advancement of Science, the Association of Women in Science, and the Institute of Electrical & Electronics Engineers (IEEE). Her awards for research and teaching awards include the NSF Presidential Young Investigator Award (1987), the Kiekhofer Distinguished Teaching Award (University of Wisconsin 1990), the American Society of Engineering Education AT&T Foundation Teaching Award (1991), the Eta Kappa Nu

C. Holmes MacDonald Distinguished Young Electrical Engineering Teaching Award (1993), the Benjamin Smith Reynolds Teaching Award (University of Wisconsin 1994), the W.M. Keck Foundation Engineering Teaching Excellence Award (1994), the ASEE George Westinghouse Award (1995), and the IEEE/HP Harriet B. Rigas Award (1995). Denton earned her BS, MS (1982), and PhD (1987) in electrical engineering at MIT and conducted research on microelectromechanical systems (MEMS) as an enabling technology particularly in life sciences applications.

BARBARA J. GROSZ is Higgins Professor of Natural Sciences in the Division of Engineering and Applied Sciences and Dean of Science of the Radcliffe Institute for Advanced Study at Harvard University. Grosz is known for her seminal contributions to the fields of natural-language processing and multi-agent systems. She developed some of the earliest and most influential computer dialogue systems and established the research field of computational modeling of discourse. Her work on models of collaboration helped establish that field of inquiry and provides the framework for several collaborative multi-agent systems and human computer interface systems. She has been elected to the American Philosophical Society and the American Academy of Arts and Sciences. A Fellow of the American Association for Artificial Intelligence, the American Association for the Advancement of Science, and the Association for Computing Machinery; recipient of the UC Berkeley Computer Science and Engineering Distinguished Alumna Award and of awards for distinguished service from major AI societies. Grosz is also widely respected for her contributions to the advancement of women in science. She chaired the Harvard Faculty of Arts and Sciences (FAS) Standing Committee on the Status of Women when it produced the report, "Women in Science at Harvard; Part I: Junior Faculty and Graduate Students" in 1991. She was Interim Associate Dean for Affirmative Action at Harvard in 1993-1994 and served on the FAS Ad Hoc Committee on Faculty Diversity from 1998-2001 and the Standing Committee on Women from 1988-1995 and again in 1999. Grosz recently chaired the 2005 Harvard Task Force on Women in Science and Engineering. Before joining the faculty at Harvard, she was Director of the Natural Language program at SRI International and co-founder of the Center for the Study of Language and Information. Grosz received an AB in Mathematics from Cornell University and a PhD in Computer Science from the University of California, Berkeley.

JO HANDELSMAN is an HHMI professor in the Department of Plant Pathology at the University of Wisconsin-Madison. She received a BS in agronomy from Cornell University and a PhD in molecular biology from University of Wisconsin-Madison. In addition, from 1997 to 1999, she was director of the Institute for Pest and Pathogen Management at University of Wisconsin-Madison. Handelsman studies the communication networks of microbial communities. She has coauthored a book about inquiry-based biology teaching entitled *Biology Brought to Life*. In 2002, she was named Clark Lecturer in Soil Biology and received the Chancellor's University Teaching Award at University of Wisconsin-Madison. In addition, she has been very active in achieving equity for women and minorities on campus, which was recognized with the Cabinet 99 Recognition Award. She contributed to the inception of the Women in Science and Engineering residence hall; has chaired the provost's Climate Working Group, an initiative dedicated to improving the campus climate for women and people of color; and, through a National Science Foundation grant, established, along with others, the Women in Science and Engineering Leadership Institute.

NANNERL O. KEOHANE is currently serving as the Laurance S. Rockefeller Distinguished Visiting Professor of Public Affairs at Princeton University. She was the eighth president of Duke University, serving from 1993-2004. Keohane came to Duke from the presidency of Wellesley College. She was the first woman to serve as Duke's president and among the first women to oversee a leading U.S. research university. Under her leadership, Duke launched major programs in fields ranging from genomics to ethics, raised more than $2 billion through the "Campaign for Duke," established the Duke University Health System and became a much more diverse and international institution. Keohane, the daughter of a Presbyterian minister, was born in Blytheville, Arkansas, and grew up in Arkansas, Texas, and South Carolina. She is a 1961 graduate of Wellesley who earned advanced degrees at Oxford University and Yale University before beginning a career as a professor of political science at Swarthmore College, the University of Pennsylvania and Stanford University. She returned to Wellesley in 1981, serving as its president for 12 years before moving to Duke.

SHIRLEY MALCOM is Head of the Directorate for Education and Human Resources Programs of the American Association for the Advancement of Science (AAAS). The directorate includes AAAS programs in education, activities for underrepresented groups, and public understanding of science and technology. Malcom serves on several boards—including the Howard Heinz Endowment, the H. John Heinz III Center for Science, Economics and the Environment, and the National Park System Advisory Board—and is an honorary trustee of the American Museum of Natural History. She serves as a Regent of Morgan State University and as a trustee of Caltech. In addition she has chaired a number of national committees addressing education reform and access to scientific and technical education, careers and literacy. Malcom is also a former trustee of the Carnegie Corporation of New York. She is a fellow of the AAAS and the American Academy of Arts and Sciences. She served on the National Science Board, the policymaking body of the National Science Foundation, from 1994 to 1998 and from 1994-2001 served on the President's Committee of Advisors on Science and Technology. Malcom received her doctorate in ecology from Pennsylvania State

University; master's degree in zoology from the University of California, Los Angeles; and bachelor's degree with distinction in zoology from the University of Washington. In addition she holds thirteen honorary degrees. In 2003 Malcom received the Public Welfare Medal of the National Academy of Sciences, the highest award given by the Academy.

GERALDINE RICHMOND is the Richard M. and Patricia H. Noyes Professor in the Department of Chemistry and Materials Science Institute at the University of Oregon. Richmond received her bachelor's degree in chemistry from Kansas State University and her PhD in chemical physics at the University of California, Berkeley. For the past 25 years her research has focused on the development and application of state-of-the-art lasers to study surface chemistry and physics. On a national level, Richmond has served and continues to serve on many science boards and advisory panels overseeing funding for science, technology, and education. Richmond has been honored with numerous national and regional awards for her research, her teaching, and her efforts in encouraging females of all ages to enter and succeed in science careers. In 2001, she was named Oregon Scientist of the Year by the Oregon Academy of Science. Richmond is a member of the Chemical Sciences Roundtable of the National Academy of Sciences, and a governor's appointee to the Oregon State Board of Higher Education for 1999-2006. She is the founder and chair of COACh (Committee on the Advancement of Women Chemists) and was the 2005 winner of the ACS Award for Encouraging Women into Careers in the Chemical Sciences.

ALICE M. RIVLIN is a Visiting Professor at the Public Policy Institute of Georgetown University and a Senior Fellow in the Economic Studies program at the Brookings Institution. She is the Director of the Greater Washington Research Program at Brookings. Before returning to Brookings, Rivlin served as Vice Chair of the Federal Reserve Board from 1996 to 1999. She was Director of the White House Office of Management and Budget from 1994 to 1996, and Deputy Director (1993-1994). She served as Chair of the District of Columbia Financial Management Assistance Authority (1998-2001). Rivlin was the founding Director of the Congressional Budget Office (1975-1983). She was director of the Economic Studies Program at Brookings (1983-1987). She also served at the Department of Health, Education and Welfare as Assistant Secretary for Planning and Evaluation (1968-1969). Rivlin received a MacArthur Foundation Prize Fellowship, taught at Harvard, George Mason, and New School Universities, has served on the Boards of Directors of several corporations, and was President of the American Economic Association. She is currently a member of the Board of Directors of BearingPoint and the Washington Post Company. She is a frequent contributor to newspapers, television, and radio, and has written numerous books. Her books include *Systematic Thinking for Social Action* (1971), *Reviving the American Dream* (1992), and *Beyond the Dot.coms* (with Robert Litan, 2001). She is co-editor (with Isabel Sawhill) of *Restoring Fiscal Sanity: How to Balance the Budget* (2004) and (with Litan) of *The Economic Payoff from the Internet Revolution* (2001). Rivlin was born in 1931 in Philadelphia, Pennsylvania and grew up in Bloomington, Indiana. She received a BA in economics from Bryn Mawr College in 1952; and in 1958 a PhD from Radcliffe College (Harvard University) in economics.

RUTH SIMMONS is president of Brown University. Simmons has created an ambitious set of initiatives designed to expand the faculty; increase financial support and resources for undergraduate, graduate, and medical students; improve facilities; renew a broad commitment to shared governance; and ensure that diversity informs every dimension of the university. These initiatives have led to a major investment of new resources in Brown's educational mission. A French professor before entering university administration, President Simmons also holds an appointment as a professor of comparative literature and of Africana studies at Brown. She graduated from Dillard University in New Orleans before completing her PhD in Romance languages and literatures at Harvard. She served in various administrative roles in the University of Southern California, Princeton University, and Spelman College before becoming president Smith College, the largest women's college in the United States. At Smith, she launched a number of initiatives including an engineering program, the first at an American women's college. Simmons is the recipient of many honors, including a Fulbright Fellowship, the 2001 President's Award from the United Negro College Fund, the 2002 Fulbright Lifetime Achievement Medal, and 2004 Eleanor Roosevelt Val-Kill Medal. She has been a featured speaker in many public venues, including the White House, the World Economic Forum, the National Press Club, the American Council on Education, and the Phi Beta Kappa Lecture at Harvard University. She has been awarded numerous honorary degrees.

ELIZABETH SPELKE is Professor of Psychology and Co-Director of the Mind, Brain, and Behavior Initiative at Harvard University. She studies the origins and nature of knowledge of objects, persons, space, and number, by assessing behavior and brain function in human infants, children, human adults and non-human animals. A member of the National Academy of Sciences and the American Academy of Arts and Sciences, and cited by *Time* magazine as one of America's Best in Science and Medicine, her honors include the Distinguished Scientific Contribution Award of the American Psychological Association and the William James Award of the American Psychological Society.

JOAN STEITZ is Sterling Professor of Molecular Biophysics and Biochemistry at Yale University School of Medicine and an investigator at the Howard Hughes Medical Institute. She earned her BS in chemistry from Antioch College in 1963, and her PhD in biochemistry and molecular biology from Harvard University in

1967. She spent the next three years in postdoctoral studies at the MRC Laboratory of Molecular Biology in Cambridge, and joined the Yale faculty in 1970. Steitz is best known for discovering and defining the function of small nuclear ribonucleoproteins (snRNPs), which occur only in higher cells and organisms. These cellular complexes play a key role in the splicing of pre-messenger RNA, the earliest product of DNA transcription. Steitz is a member of the National Academy of Sciences, the American Association of Arts and Sciences, the American Philosphical Society, and the Institute of Medicine. She is a recipient of the National Medal of Science, 11 honorary degrees, and a Gairdner Foundation International Award, among others. She serves on numerous review and editorial boards.

ELAINE WEYUKER is a principal technical staff member at AT&T Labs at Florham Park, NJ. Weyuker received a PhD in Computer Science from Rutgers University, and an MSE from the Moore School of Electrical Engineering, University of Pennsylvania. Before moving to AT&T Labs in 1993, she was a professor of Computer Science at the Courant Institute of Mathematical Sciences of New York University, NY, where she had been on the faculty since 1977. Her research interests are in software engineering, particularly software testing and reliability, and software metrics, and has published many papers in those areas. Among her honors, she has been elected to the National Academy of Engineering and as a Fellow of the IEEE, and has been named a Fellow of the ACM (Association of Computing Machinery). Weyuker is one of only two female AT&T Fellows. In each of the past 6 years, The Journal of Systems and Software has rated her as one of the top five software engineering researchers in the world. In November 2001, the NYC YWCA honored Weyuker as a "Woman Achiever" for both her career achievements and her community service. She has made major contributions to the formal foundations of testing and to establishing testing as an empirical discipline, and has been a prime mover in making testing a recognized, professional specialty. She has been a lecturer, teacher, and mentor, and has been actively involved in professional activities. She was a founding member of the ACM Committee on the Status of Women and Minorities, which was established to improve the status of under-represented groups by developing programs to target girls and young minority members. During her tenure, the committee established a successful distributed mentoring program.

MARIA T. ZUBER is the E.A. Griswold Professor of Geophysics at the Massachusetts Institute of Technology where she also leads the Department of Earth, Atmospheric, and Planetary Sciences. Zuber has been involved in more than half a dozen NASA planetary missions aimed at mapping the Moons, Mars, and several asteroids. She received her BA from the University of Pennsylvania and ScM and PhD from Brown University. She was on the faculty at Johns Hopkins University and served as a research scientist at Goddard Space Flight

Center in Maryland. She is a member of the National Academy of Sciences and American Philosophical Society, and a fellow of the American Academy of Arts and Sciences and of the American Geophysical Union, where she served as president of the Planetary Sciences Section. Among her awards are the NASA Distinguished Public Service Medal, the NASA Scientific Achievement Medal, and Brown University Horace Mann Medal, as well as a Scientific Achievement Award from the American Institute of Aeronautics and Astronautics. Professor Zuber served on the Mars Program Independent Assessment Team that investigated the Mars mission losses in 1999, and more recently on the Presidential Commission on the Implementation of the United Space Exploration Policy tasked with conceiving a plan to implement President Bush's Vision for Space Exploration. In 2002, *Discover* magazine named her one of the 50 most important women in science.

APPENDIX D
STATEMENT OF TASK

Research in science and engineering has been and remains central to the US role in the world, the culture of the nation, its continuing economic development, and its security. It is imperative that the nation access its entire talent pool. However, it is clear from several recent studies that while women are an increasing proportion of those earning undergraduate and graduate degrees in science and engineering fields, they have not been hired into academic positions commensurate with this increasing representation. Ultimately, this means that the academic research enterprise is missing out on talent, and will under perform relative to its potential.

The study committee will integrate the wealth of data available on gender issues across all fields of science and engineering. The committee will focus on academe, but will examine other research sectors to determine if there are effective practices in place relevant to recruiting, hiring, promotion, and retention of women science and engineering researchers. Throughout the report, profiles of effective practices, scenarios, and summary boxes will be used to reinforce the key concepts.

The committee is charged to:

(1) Review and assess the research on gender issues in science and engineering, including innate differences in cognition, implicit bias, and faculty diversity.
(2) Examine the institutional culture and practices in academic institutions that contribute to and discourage talented individuals from realizing their full potential as scientists and engineers.
(3) Determine effective practices to ensure women doctorates have access to a wide range of career opportunities, in academe and in other research settings.
(4) Determine effective practices on recruiting and retention of women scientists and engineers in faculty positions.
(5) Develop findings and provide recommendations based on these data and other information the committee gathers to guide the following groups on how to maximize the potential of women science and engineering researchers:
    (a) Faculty: roles in hiring, promotion, retention, and mentoring
    (b) Deans and Department Chairs: roles in hiring and promotion and equitable provision of resources
    (c) Academic Leadership: roles in hiring, promotion, resource allocation, tracking, and setting the tone for institutional culture
    (d) Funding Organizations: roles in education and training, compensation levels, review, and tracking of grant applicant and recipient data.
    (e) Government: roles in enhancing and diversifying access to education, training, and research funding, and in ensuring that data about program users are collected and available for assessment purposes.